Headquarters
Department of the Army
Washington, DC, 13 July 1992

Engineer Operations Short of War

Contents

Page

Page

Preface

SCOPE

FM 5–114 contains doctrine for the conduct of engineer operations short of war. These are operations conducted in peacetime or during conflict. The manual addresses the impact of political, informational, and economic factors on military planning in these environments. It includes chapters relating to each category of operations short of war and addresses engineer–specific missions within each category. The chapters outline planning considerations, coordination requirements, and key operational concepts. Where possible, regulatory constraints are explained; however, these constraints may change as new laws are passed. Appendices in this manual provide specific information on force protection, unit deployment considerations, the capabilities of selected engineer cellular teams, and the United States (US) Army Corps of Engineers (USACE). Throughout this manual "military engineer" means a member of an Army troop unit; "USACE" or "USACE engineer" means either a military or civilian member of the major Army command (MACOM) USACE; and "US Army engineer" means both types.

PURPOSE

FM 5–114 provides doctrine to commanders and staffs at battalion level and above concerning the employment of engineer units at all levels in military operations short of war. It identifies related references, highlights key concepts, and provides planning guidance to engineer units preparing to conduct operations short of war. Other branches will use the manual for guidance when employing engineer units in this arena. The manual applies to both Active Component (AC) and Reserve Component (RC) units.

FM 5–114 serves as a source of information for non–Department of the Army (DA) planners who may be involved in operations with Army engineer units, to include members of other Department of Defense (DOD) agencies; non–DOD agencies such as the Department of State and the Drug Enforcement Administration (DEA); and federal, state, and local emergency management agencies.

The proponent for this publication is HQ, TRADOC. Submit changes for improving this publication on DA Form 2028 (Recommended Changes to Publications and Blanks Forms) and forward it to Commandant, US Army Engineer School, ATTN: ATSE–TDM–P, Fort Leonard Wood, MO 65473–5000.

Unless otherwise stated, masculine nouns and pronouns do not refer exclusively to men.

Chapter 1

The Strategic Environment

National Strategy Goals—Throughout our history, our national security strategy has pursued broad, consistent goals. We have always sought to protect the safety of our nation, its citizens, and its way of life. We have also worked to advance the welfare of our people by contributing to an international environment of peace, freedom, and progress in which our democracy-and other free nations—can flourish.

National Security Strategy of the United States
The President of the United States
March 1990

Military operations short of war are of increasing importance to the United States (US). These are operations conducted in peacetime or during conflict. Nearly all armed conflicts of the past 40 years have occurred in the Third World. These are the diverse, developing countries primarily situated in Asia, the Middle East, Africa, Latin America, and the Caribbean Sea. During this 40–year period, all conflicts in which the US has been involved (directly or indirectly) have occurred in this region. Continuing changes in the world geopolitical scene have increased the chance that future US military operations will fall within the arena of operations short of war.

THE OPERATIONAL CONTINUUM

Each nation has various means by which it can achieve its national goals. These include the political, economic, informational, and military elements of national power. The political element of power consists of moral and political example, alliance relationships, public diplomacy, and diplomatic mediation. Methods by which a nation employs its economic power include economic incentives and sanctions, developmental assistance, and financial aid. The informational element includes cooperation in the areas of science and technology, participation in international organizations, and provision of information. Military strength and security assistance to other nations are two facets of the military element. In most situations, a nation exercises all four elements of national power to varying degrees. For example, US participation in an alliance such as the North Atlantic Treaty Organization (NATO) uses all four elements to achieve national aims. In describing environments in which US military forces may conduct operations, it is vital to understand the interaction of these elements.

The operational continuum is the range of environments in which the US military conducts operations. The continuum consists of three general states: peacetime, conflict, and war. The state of war is a violent environment in which the national survival of a belligerent is at stake. In war, military force contributes directly to the achievement of strategic aims. It seeks to create conditions that make it possible to achieve the desired strategic end state. The political, economic, and informational elements of national power support the military effort to establish these conditions.

In the less violent environments of the operational continuum, US forces operate under conditions of peace or conflict. "Short of war" is a collective term for these states of the environment. A key concept relating to these states is low intensity conflict (LIC). This is a political–military confrontation between contending states or groups below conventional war and above the routine, peaceful competition among states. LIC often involves protracted struggles of competing principles and ideologies. It ranges from subversion to the use of armed force. LIC is waged by a combination of means, employing political, economic, informational, and military elements. Low intensity conflicts are often localized. They generally occur in the Third World but have regional and global security implications. In LIC, the contribution of military force to the US strategic aim is indirect. Military operations support nonmilitary actions that create conditions under which strategic aim can be realized. Frequently, these actions aid friendly governments or resistance groups.

The key difference between a LIC environment and war is the way the elements of national power are employed. In war, the US directly employs the military element to establish conditions that enable the nation to achieve its strategic aims. The other elements of national power support the military element. In a LIC environment, the US directly employs the political, economic, and informational elements to establish conditions that help secure US strategic aims. Ideally, the military element directly supports the other elements. An example is a show of force designed to express US support for a host nation's (HN's) government. This is not a hostile act, but it shows US resolve and supports political initiatives. Sometimes the direct use of military force maybe necessary, but this is not the preferred approach to achieving national goals.

It is essential to understand that the "low intensity" description within the term LIC is a US perspective. LIC involves threats that are usually subtle and indirect and that occur over an extended period of time but which also have serious implications for US national security interests. From the perspective of the government of a country where a "low intensity" conflict is occurring, the intensity may be very high. The conflict may threaten the country's existence. To US military personnel involved in an operation in a LIC environment, the perceived level of intensity from the national perspective is unimportant. To those individuals the environment is one of conflict.

Within an area of operations, more than one environment may exist at the same time. For example, within a country where US forces are operating, there maybe both a limited war and a LIC environment. Within a unified Commander in Chiefs (CINC's) area of responsibility, environments may range from peacetime to war simultaneously. Commanders must be aware of the specific environment in which they are operating. The environment affects such planning considerations as force protection, rules of engagement (ROE), and interaction with the local populace.

Escalation of conflict between states of the continuum may occur, but it is not inevitable. For example, LIC does not necessarily lead to war. The objective of operations within LIC, in fact, should be to resolve conflict on terms favorable to US interests without resorting to war. On the other hand, war does not always resolve all aspects of contention between nations. A war may evolve into some form of LIC. Military forces must be prepared to adjust their operations as these transitions occur.

THE ARMY IN PEACETIME

Under peacetime conditions, the political, economic, and informational elements of the US are the primary mechanisms for achieving national goals. Those who direct the agencies that administer these elements take the lead in establishing policies and programs that support the country's objectives. The military assets of the US serve two purposes in a peacetime environment.

During peacetime, the primary mission of the military element is to deter conflict and war. A trained and ready force reduces the chance that operations in more hostile environments will occur, Two conditions must be met for the US armed forces to accomplish this mission. First, the forces must be perceived as being capable of responding to any threat or contingency that arises. This capability is linked to the levels of manning and training, the material status, and the deployability of the forces. Second, the international community must perceive that the US government will apply the military element. These two conditions equate to deterrence.

During peacetime, the secondary mission of the military element is to support political, economic, and informational efforts to achieve US goals. Military operations in this environment provide an indirect application of the military element of power, compared with the direct application of the military element in the environment of war. Activities by armed forces under peacetime conditions must be closely coordinated with the agencies responsible for directing the use of the other elements of power. This is necessary to ensure unity of effort toward achieving US goals and a consistent approach by all US agencies in dealing with members of the international community.

A great percentage of military operations conducted in the pursuit of peace will occur outside the US. These operations will generally be conducted to support another US government agency's programs. Often the goal of these programs will be to promote stability within a nation or region of the world. In this context, stability is not a status quo but orderly progress for national or regional development.

Within the military establishment, engineer forces are particularly well suited to conduct operations in this environment. The capabilities, flexibility, and generally nonthreatening nature of military engineer units readily support the types of programs that are established during peacetime to achieve US national goals. Some activities in which military engineers may be involved include staff assistance visits, personnel exchanges, civic action projects, and unit deployments for training. Military engineer units may conduct these operations independently; they may be augmented by combat, combat support, combat service support, or United States Army Corps of Engineers (USACE) (civilian or military) assets; or they maybe part of a combined or joint force.

Although most military peacetime operations occur outside the US, there are several noteworthy exceptions of particular interest to Army engineers. These are disaster preparedness and relief operations, military construction (MILCON), civil works projects, topographic engineering projects, and community civic action projects. These operations do not have a direct impact on the international environment, but they are critical peacetime tasks.

LOW INTENSITY CONFLICT

There are numerous factors that contribute to the development of a LIC environment in a country or region. Among these are change, poverty, and discontent. These circumstances are often interrelated, particularly in developing nations of the Third World. The failure of political and social institutions to incorporate the general populace into the process of modernization and development may cause discontent within a society. By its very nature, modernization makes the people aware of the disparity between the haves and have-nots. This leads to rising expectations that the government of the developing nation may not be able to fulfill. The inability or unwillingness of a developing nation to meet the real or perceived needs of its people provides a fertile ground for unrest within the society. Groups within that nation, as well as other governments, may exploit such unrest.

Since the end of World War II, a host of groups and states have pursued their interests in the LIC environment. International wars and insurgencies have taken a heavy toll on lives and resources. Most have occurred in the Third World and have changed the international environment. Many Third World conflicts originated in the struggle to end the system of European colonialism. As nations were created, clashes occurred as relationships were established or conflicts developed among newly independent states. In many cases, insurgents sought to alter the political, social, and economic organization of these states, bringing about internal conflicts. The proliferation within these regions of chemical and biological weapons, and to some degree nuclear weapons, has increased the lethality of the environment. These weapons have greatly increased the ability of small states and terrorist groups to influence larger and traditionally more powerful countries.

The rapid growth of the international, illegal drug trade also contributes to instability within the world community. Drug traffickers are using increasingly sophisticated communications, transportation, and support systems in their operations. They are also using more lethal weapons in their confrontations with authorities. In many cases, the equipment used by the drug traffickers is more modern than that used by the governmental forces combatting them. Together, these factors suggest a high potential for violent conflict in the world. The current conditions that exist in the Third World suggest that future confrontations will most likely occur in that region and in the context of a LIC environment.

The most significant threats to US interests in the LIC environment do not come from individual incidents of insurgency, economic instability, drug trafficking, or isolated acts of terrorism. Rather, they result from the accumulation of unfavorable outcomes from such activities. Failure in the LIC environment can lead to the following:

- Isolation of the US from its allies, its global trading partners, and the world community.
- Loss of US access to strategic mineral and energy supplies.
- Loss of US military basing, transit, and access rights.
- Loss of international support in the counterdrug war.

- Movement of US friends and allies to positions of accommodation with hostile groups.
- Long-term gains for US adversaries.

Successful operations in a LIC environment can advance US international goals, such as the growth of freedom, democratic institutions, and free-market economies.

Success in the LIC environment requires planning and conducting operations based on the following set of imperatives:

- Primacy of the political element.
- Unity of effort.
- Adaptability.
- Legitimacy.
- Perseverance.
- Restricted use of force.

During operations in a LIC environment, political objectives drive military decisions at every level from the strategic to the tactical. Commanders and their staff officers must understand the specific political objectives and the impact of military operations on them. They must adopt courses of action that legally support those objectives even if the courses of action appear to be outside what traditional military doctrine encompasses. Direct involvement by National Command Authorities (NCA) is likely in this environment.

Unity of effort dictates that military leaders must integrate their efforts with other governmental agencies to gain a mutual advantage in the LIC environment. Commanders must consider how their plans contribute to initiatives that are also political, economic, and informational in nature. Unity of effort requires interagency integration and coordination to permit effective action within the framework of the US government.

Adaptability is the skill and willingness to change or modify existing structures and methods to accommodate different situations (for example, tactics, techniques, procedures, training, and leadership). It requires careful mission analysis, comprehensive intelligence, and regional expertise. Adaptability is more than tailoring or flexibility, which imply the use of standard techniques or structures for a variety of cases. Successful operations in a LIC environment will require the armed forces to adapt existing methods and to develop new ones appropriate to each situation.

Legitimacy is the willing acceptance by a people of the right of their government to govern or of a group or agency to make and enforce decisions. It also includes acceptance of the presence and actions of representatives of US government agencies (to include military forces) in a country. Legitimacy is not tangible nor easily quantifiable. Popular votes do not always confer or reflect real legitimacy. Legitimacy derives from the perception that authority is genuine and effective and uses proper agencies for reasonable purposes. No group or force can decree legitimacy for itself, but it can create and sustain legitimacy by its actions. Legitimacy is the central concern of all parties directly involved in a conflict.

Perseverance is the patient, resolute, and persistent pursuit of national goals and objectives. Low intensity conflicts rarely have a clear beginning or end marked by decisive actions culminating in victory. By their nature, they are protracted struggles. Even those short, sharp contingency encounters which do occur must be assessed in the context of their contribution to long–term objectives. Perseverance does not preclude taking decisive action. In fact, it requires every effort to gain and maintain social, economic, political, and military initiatives. Perseverance is needed for both civilian and military leadership to reject limited, short-term goals.

Restricted use of force refers to the judicious, prudent, and thoughtful selection and employment of forces most suitable to the mission. Restricted use of force does not preclude the possibility of applying massive or overwhelming force, when appropriate, to display US resolve and commitment. The ROE in the LIC environment will usually be more restrictive, detailed, and subject to political scrutiny than during other types of conflict. As a result, this environment is often marked by constraints on weaponry, tact its, and the level of violence. Excessive violence can adversely affect efforts to gain or maintain legitimacy and impede the attainment of both short–term and long–term goals.

All US operations within the LIC environment must be planned with due consideration for the people of the area in which the action is occurring. In countries where a representative government now exists, US operations must strengthen the image of the national and local governments in the eyes of the populace. This may not be the primary objective of any given operation, but it must be a secondary one. It is essential that the perception of the host government's strength and independence be reinforced.

A perception that the US is imposing its will in an area rather than cooperating with the nation's government is unacceptable. That perception will weaken that government and possibly worsen the conditions that caused the LIC environment. Operations conducted without regard for this aspect of LIC may meet with short–term success but will have a negative impact on the long–term interests of the US. As an example, take a friendly nation threatened by instability, unrest, or violence. Security assistance is one example of a military tool for this environment. US security assistance efforts should aid that nation's military in developing the capability to provide security for its citizens and government.

The US will employ combat operations in a LIC environment in exceptional circumstances when it cannot protect its national interests by other means. When a US response is called for, it must be according to the principles of international and domestic law. These principles affirm the inherent right of states to use force in individual or collective self-defense.

There are four broad categories of operations in the LIC environment. These are support for insurgency and counterinsurgency, combatting terrorism, peacekeeping operations (PKOs), and contingency operations (CONOPs). A US force operating in a LIC environment may be engaged in one or more of these operations at the same time. For example, a unit operating in the Middle East as part of a peacekeeping force may become the target of terrorist activities. Commanders must ensure that their planning covers the entire spectrum of the threat that they face. Detailed, accurate, and current intelligence is essential for maintaining an up-to-date threat analysis.

SUPPORT FOR INSURGENCIES AND COUNTERINSURGENCIES

An insurgency is an armed, organized political struggle. Its goals may be the seizure of power through revolutionary takeover and the replacement of an existing government. In other cases, an insurgency may only intend to extract limited political concessions that are deemed or considered unattainable through less violent means. For example, the objective of an insurgency may be to free an area from governmental control and establish an autonomous state within traditional ethnic or territorial boundaries. Insurgency and counterinsurgency are two aspects of the same process. Insurgency focuses on change in political control and requires the extensive use of covert instruments and methods. Counterinsurgency, on the other hand, uses mainly overt methods. It includes promises of orderly change within the existing system. Because of these differences, supporting doctrine varies for insurgency and counterinsurgency.

Since there is a political goal in all insurgences and counterinsurgencies, political leadership is of paramount importance. Military power plays a supporting role. The leadership of the US may choose to assist in either countering or supporting an insurgency. This will depend on the aims of the insurgents. To effectively operate in this arena, US leaders must identify the root causes of an insurgency. They must then assess the current situation, estimate the insurgency's near–term course, and carry out measures to either support or thwart it. The better a decision maker understands the motivations of the contenders in an insurgency, the more likely he will be able to anticipate and influence their behavior. Whether the US government supports or opposes an insurgency, knowledge of the nature of the conflict and the cultural and geographic environment is essential for determining appropriate actions.

Insurgencies succeed by mobilizing human and material resources to provide both active and passive support to the insurgent movement. These resources provide skilled workers and willing fighters, raise money, and acquire weapons, intelligence, and supplies of all kinds. The mobilization of an insurgent movement may stem from a growing popular dissatisfaction with existing political, social, economic, ethnic, cultural, or religious conditions. The insurgent leadership broadcasts these dissatisfactions and attempts to place the blame for the grievances on the existing government. The insurgents attempt to convince the people that the government cannot or will not act to remedy the problems. At the same time, the insurgents may offer a program or proposal to improve conditions. A government's inability or unwillingness to exercise economic foresight, to recognize the magnitude of the people's grievances, and to effectively coordinate activities may inhibit social progress. This can increase the risk of insurgent exploitation.

An insurgency attempts to exploit actual or perceived governmental weaknesses, such as failure to maintain law and order, or overreaction to civil disturbances. Critical to the insurgent movement is the need to form a support base among the population, regardless of whether support is voluntary or provided by force or coercion. The key requirement for insurgents is to provide organizational and management skills to transform disaffected people into an effective political element. Insurgencies calling for the overthrow of the government incorporate a revolutionary doctrine and use armed force as an element of policy. The insurgents vie with the existing government for political legitimacy and the right to lead and represent the people.

Support to insurgencies (resistance movements) is termed a special activity within US regulatory and statutory guidelines. Although support to resistance movements can be conducted overtly or with low visibility, much of this support must be provided covertly. Such action may require the use of clandestine techniques. These include grey and black propaganda, secret funding, deception, paramilitary operations, and covert logistical networks. US armed forces will aid an insurgency as part of an interagency effort. Military support will primarily consist of training and advising insurgent forces in the tactics, techniques, and procedures of unconventional warfare (UW). In this context, UW is intended to be a protracted politico-military activity that seeks to weaken the control and legitimacy of the established government or occupying power. At the same time, it attempts to increase the control and legitimacy of the resistance movement.

The US supports HN counterinsurgency efforts based on the concept of internal defense and development (IDAD). This concept assumes a HN government is responsible for developing and executing programs to prevent or defeat subversion, lawlessness, or insurgency. The host government must identify the genuine grievances of its people and take political, economic, and social actions to redress them. Carefully planned and implemented and properly publicized development programs can serve the interests of population groups and deny exploitable issues to insurgents. IDAD is a strategy to prevent subversion, lawlessness, or insurgency and if such threats should develop, to combat them.

The US uses its military resources to provide support to an HN's counterinsurgency operations in the context of foreign internal defense (FID). FID is participation by US civilian and military agencies in another nation's action programs to free and protect that nation's society from subversion, lawlessness, and insurgency. FID operations are US efforts to support a friendly government facing a threat to its internal stability and security. The US armed forces can provide resources such as material, advisors, and trainers to support HN counterinsurgency operations. The US may provide more direct forms of armed support when it is in the security interest of the US and the HN requests it.

All US FID resources must be coordinated with the country team in the US diplomatic mission. This is to ensure that only those resources appropriate to local conditions are employed to achieve HN IDAD goals. FID programs should enhance the capability of indigenous military and paramilitary forces to perform unilateral internal defense missions. Military civic-action programs are a subset of FID. These programs use predominantly indigenous military forces on projects useful to the local population at all levels. They support such areas as education, training, public works, agriculture, transportation, health, and sanitation. The programs also serve to improve the standing of the military forces with the population. This serves to increase the legitimacy of the government while alleviating some conditions that cause hardship and dissatisfaction among the people.

COMBATTING TERRORISM

The Department of Defense (DOD) defines terrorism as the calculated use of violence or threat of violence to inculcate fear; intended to coerce or intimidate governments or societies in the pursuit of goals that are generally political, religious, or ideological. Terrorists' objectives may range from the advancement of separatist causes to gaining revenge for ethnic grievances to fundamental social and political revolution. Unstated objectives may include provoking overreaction by their adversary or stimulating a response that acknowledges the terrorists' power and legitimacy. Their methods may be equally diverse, ranging from planting explosive devices in public places to carrying out kidnappings and assassinations. Other techniques include hijacking, arson, hoaxes, and armed attacks. Regardless of their approach, the goal of all terrorists is to impose their will on society by using violence to create an atmosphere of fear. Terrorists want people to feel defenseless and doubt their government's capacity to protect them. This in turn undermines the legitimacy of the government, its policies, or both.

The key feature that distinguishes terrorism from other forms of political violence is the willingness of terrorists to attack "innocent" targets. International law restrains the use of armed forces to minimize harm to noncombatants. The same laws apply to recognized insurgencies. The terrorist has no such restraints and often attacks people with no connection to his cause or issue. This is why terrorist acts are by definition illegal in conditions of peace, conflict, or war. This is also why, by committing illegal acts of violence, insurgents can become terrorists.

Terrorist groups are distinguished primarily by governmental affiliation. This helps security planners anticipate terrorists targets and their degree of sophistication in intelligence and weaponry. Terrorists are generally classified as nonstate-supported, state-supported, or state-directed. However, networking and mutual support make any categorization somewhat arbitrary. Nonstate-supported terrorist groups operate autonomously, receiving no significant support from any government. Italy's Red Brigades and the Basque Euskadi ta Askatasuna are examples of such nonstate-supported groups. State-supported terrorist groups generally operate independently but receive support from one or more governments. The Popular Front for the Liberation of Palestine is one example of this type of group. State-directed terrorist groups operate as an agent of a government. They receive intelligence, logistics, and operational support from that government. Libyan "hit teams" targeted against Libyan exiles are an example of state-directed terrorists. Terrorist events may be classified by their immediate objectives. Five types of terrorist objectives are—

- Recognition.
- Coercion.
- Intimidation.
- Provocation.
- Insurgency support.

National or international recognition of a cause may be the objective of a terrorist campaign at its outset. The reasons for seeking recognition may include attracting recruits, obtaining funds, or showing strength. Groups seeking this recognition use incidents that are likely to attract media attention. These include hijacking an aircraft, kidnapping prominent people, seizing occupied buildings, or hostage-barricade incidents.

Coercion is the attempt to force a desired behavior by persons, groups, or governments. This objective calls for a strategy of very selective targeting. It may rely on publicly announced bombings, destruction of property, and other acts that are less violent than taking human life.

Intimidation differs from coercion. Intimidation attempts to prevent persons or groups from acting. Terrorists may use intimidation to reduce the effectiveness of security forces by making them afraid to act. Intimidation can discourage competent citizens from seeking or accepting positions within the government. The threat of violence can also keep the general public from taking part in important political activities such as voting. As with coercion, terrorists use a strategy of selective targeting.

Provoking overreaction by government forces may be the objective of terrorist acts. This strategy calls for attacking targets symbolic of the government, such as the police, military, or other officials. Attacks of this type show vulnerability to terrorist acts and contribute to a loss of confidence in the government's ability to provide security. More importantly, if the security forces' response is heavy-handed, the resulting oppression can create public sympathy, passive acceptance, or active support for the terrorist group.

Terrorism in support of an insurgency is likely to include provocation, intimidation, coercion, and the quest for recognition. Terrorism can also aid an insurgency by causing the government to overextend itself in attempting to protect all possible targets. Other uses of terrorist skills in insurgencies include acquiring funds, coercing recruits, obtaining logistical support, and enforcing internal discipline. Use of these techniques must be weighed against the impact on attempts to gain popular support for the insurgency.

A terrorist organization may pursue one, some, or all of these objectives. The terrorist organization may establish its objectives and strategy, or the government supporting the terrorist organization may direct them. In either case, the military planner must identify these objectives and strategies to defeat the terrorist organization.

Terrorism can be international in scope and in some instances is aided and abetted by state sponsors. Therefore, the threat posed to US citizens and security interests abroad may require a US-armed-forces response. This response can occur in two ways: deterring acts of terrorism through active and passive measures (antiterrorism (AT)) or employing forces to directly combat terrorists (counterterrorism (CT)).

AT is the range of defensive measures used to reduce the vulnerability of persons and property to terrorism. The basics of AT include collecting and disseminating timely threat information, conducting terrorism awareness information programs, and using sound defensive measures. Because absolute protection against terrorist activities is not possible, protective plans and procedures are based on the threat. Protective measures should strike a reasonable balance between the protection desired, mission requirements, and available manpower and financial resources.

CT is the full range of offensive measures taken to prevent, deter, and respond to terrorism. CT is based on extensive preparation and planning. CT operations normally require specially trained personnel capable of mounting swift and effective action. Response measures can include preemptive, retaliatory, and rescue operations. The type of forces and the command and control relations used in CT operations depend on the location, type of incident, and degree of force required. Force selection criteria are normally governed by legal and political restraints. In planning and conducting CT operations, it is essential that realistic and sensible ROE be prepared to minimize civilian and/or collateral damage. ROE maybe developed by the NCA with the advice of their military and civilian advisors.

PEACEKEEPING OPERATIONS

PKOs are military operations conducted with the consent of the belligerent parties in a conflict. The objective of PKOs is to maintain a negotiated truce and to facilitate diplomatic resolution of the conflict between the belligerents. The US may participate in PKOs sponsored by the United Nations (UN) or another organization, in cooperation with other countries, or unilaterally. US personnel may function as impartial observers, as part of an international peacekeeping force, or in a supervisory and assistance role. The operations may take many forms. Possible PKOs include supervision of—

- Cease-fires.
- Withdrawals and disengagements.
- Prisoner of war (POW) exchanges.
- Demilitarization and demobilization.

Peacekeeping often involves ambiguous situations requiring the peacekeeping force to deal with extreme tension and violence without becoming a participant. Peacekeeping operations follow diplomatic negotiations that establish the mandate for the peacekeeping force. The mandate describes the scope of PKOs in detail. It often states the size and type of force each participating nation will contribute. The mandate should clearly state the following:

- The terms or conditions the HN intends to impose on the presence of the force or mission.
- A clear statement of the rights and immunities of force or mission members under jurisdiction of the international agency.
- A clear statement of the functions the peacekeeping force is to perform.

Based on the mandate, diplomats will then establish stationing agreements with the advice of the commander concerned. These are often referred to as Status of Forces Agreements (SOFAs). These agreements between the HNs, sponsor, and contributors will establish the detailed legal status of the force or mission. Based on the peacekeeping mandate and the stationing agreement, the specific terms of reference, follow-on command directives, and ROE are established. The terms of reference describe how the US will implement its portion of the PKO. They include the-

- Mission.
- Command relationships.
- Organization.
- Logistics support.
- Accounting procedures.
- Responsibilities of the US contingent to the peacekeeping force.
- Coordination and liaison arrangements.

They may also include public affairs procedures and any bilateral relationship with other national contingents, in addition to those described in the mandate.

PKOs generally have three levels (or tiers) of organization: the political council, the military peacekeeping command, and the military area command. The term "peacekeeping force" includes all three levels. The political council is the highest level of the peacekeeping

organization. It provides a mechanism for negotiating and coordinating with leaders of disputing parties. Through negotiation, the council encourages self–sustaining solutions that are acceptable to disputing factions. The political council receives the mandate for the PKO and coordinates that status of forces with the belligerents. The chief of the peacekeeping force may be a member of the political council.

Overall control of peacekeeping forces exists at the military peacekeeping command level. Control and staffing at this level is normally multinational. The force commander exercises operational control of the combined forces, with command functions remaining within national channels. The military peacekeeping command rarely has the authority to negotiate political matters. The missions of the command include-

- Deterring violent acts by the disputants.
- Protecting vital installations and critical facilities.
- Informing the political council of peacekeeping force requirements.
- Collecting and providing information to the political council.
- Ensuring impartiality of peacekeeping forces.

The command issues directives and instructions concerning operations and procedures.

The operating level of the peacekeeping force is the military area command. The area command usually consists of forces from a single nation. It operates in a specific area of responsibility and reports to the military peacekeeping command. It may receive logistic support from the peacekeeping command or from its own national channels. The area command places distinctive markings on all uniforms and equipment to ensure that they are clearly identified as members of the peacekeeping force. The area command deters violent acts by its physical presence at locations prone to violence. It collects information through overt means such as observation posts, patrols, aerial reconnaissance, and conversations with local inhabitants.

The following eight principles are fundamental to PKOs. Some of them are discussed in the following paragraphs. For additional information on these principles and the other principles of peacekeeping, see Field Manual (FM) 100–20.

- Consent.
- Neutrality.
- Balance.
- Single manager control.
- Concurrent action.
- Unqualified sponsor support.
- Freedom of movement.
- Self-defense.

The degree of consent of all parties involved in a PKO has a major impact on its success. The disputing parties show their desire for this operation by their degree of consent. Nations participating in the peacekeeping force must do so without reservation or restriction. They must fully consent to PKOs, whether out of self–interest or for purely humanitarian reasons. Consent also applies to other interested states. They may support PKOs or at least refrain from impeding them.

Organizations or countries contributing to a PKO should give the peacekeeping force their full and unqualified support according to the terms of the mandate establishing the force. This support may be logistical, financial, or political. The contributing groups should permit the peacekeeping force to operate freely, within policy guidance, but without unnecessary interference.

The use of force in self-defense is essential to the concept of PKOs. Self-defense is an inherent right; it is the one principle that cannot be affected by consent. ROE describe circumstances and the manner in which peace keepers may use force to resist attempts to prevent them from doing their duties. ROE should be clearly and unambiguously stated in the mandate.

PKOs require commanders to position their units in potentially hostile environments. To be effective and maintain their security, the peacekeeping force must remain impartial. The commander should withdraw his force if conditions deteriorate and jeopardize the force's impartiality. Controlling violence in PKOs requires a combination of techniques. These include observing, patrolling, investigating complaints, negotiating, and mediating.

CONTINGENCY OPERATIONS

CONOPs are politically sensitive military operations normally characterized by short–term, rapid projection or employment of forces in conditions short of war. They are often undertaken in crisis-avoidance or crisis-management situations requiring the use of military elements to enforce or support diplomatic and informational initiatives. CONOPs include but are not limited to the following missions:

- Disaster relief outside the US.
- Counterdrug operations.
- Security assistance surges.
- Noncombatant evacuation operations (NEOs).
- Rescue and recovery operations.
- Shows of force and demonstrations.
- Operations to restore order.
- Strikes and raids.

CONOPs involve tailored forces and are usually joint or combined in scope. Planners must establish clear command relationships and communications procedures because the lead organization varies according to the type of operation. Military forces employed in CONOPs will normally use service-specific tactical doctrine or joint tactics, techniques, and procedures to execute their mission. The forces employed will be chosen from designated contingency forces that have planned and trained to the required standard necessary for the rapid and successful conduct of the operation.

The three principles critical to CONOPs are coordination, balance, and planning for uncertainty. Coordination is essential for success due to the nature of the forces involved and the fact that the operations are being conducted to support political aims. Coordination must occur among the forces that will actually execute the operation and between military leaders and the lead and supporting federal agencies involved.

A balance must be struck between overriding political goals and the scale, intensity, and nature of military operations supporting those goals. A unit commander must provide for the security of his force within the constraints of the ROE and the political sensitivity of the situation. The commander requires clearly stated objectives and operational parameters to balance his security needs with national policies.

Situations filled with uncertainty require detailed but flexible planning that incorporates the principles of coordination and balance. This requires a full awareness of the political and social realities of the area in dispute. In preparing for CONOPs, operational–, logistical-, engineer-, and intelligence-support planning must be comprehensive.

CONOPs have a strong psychological impact on the attitudes and behavior of domestic and foreign audiences. The operations must be planned with this in mind. As an example, attacks on terrorist sanctuaries and shows of force show US resolve to isolate and/or punish violations of international law and human rights. However, these operations must be planned and executed with consideration for their impact on public opinion as well as their direct impact on the people living in the area of operations. In this environment, a military operation must be militarily, politically, and informationally successful to achieve its overall goal.

For a more detailed explanation of responsibilities, relationships, and operations in the LIC environment, refer to FM 100–20.

Chapter 2

Engineer Peacetime Activities

The Military Engineer in Peace—The military engineer should be made an effective instrument for war but he should also be utilized to the limit of his useful capacity in time of peace. There is no conflict of requirements. On the contrary such duty is, in fact, necessary to his efficiency in war. There is a strong family likeness in all engineering and there are few engineering activities in war which are not analogous to the activities in peace. There is no reason why the military engineer, whose primary function is war, should not also be utilized on peace-time engineering work during the intervals between wars, and it is necessary to his efficiency for war that he be given this experience in peace.

"Army Construction and the Corps of Engineers"
COL G.A. Youngberg and COL C.O. Sherrill
The Military Engineer, March-April 1920

In peacetime, the military element of power supports the other three elements in achieving United States (US) national goals. The actions of US Army engineers can enhance the domestic and foreign programs of other US government agencies. Army engineer activities support US security objectives worldwide. They promote a favorable image of the US and its democratic way of life. US Army engineers operating in foreign nations improve relations between the US and host governments. These engineers teach basic skills to indigenous civilian and military personnel. They work together on projects that support host–nation (HN) institutional and infrastructure development. These projects serve as training opportunities for both US and HN personnel. In addition, the projects provide social, economic, or other benefits to the HN. This chapter describes the coordination of US assistance activities within an HN. It explains several programs that increase the expertise of local military engineers and associated civilians (transportation planners, civil engineers, and so forth). These programs transfer skills through HN interaction with US personnel. US Army engineers may be asked for input on these programs as part of an overall developmental strategy for a country.

NATION ASSISTANCE

Nation assistance is not a separate category of activity. It includes all cooperative actions taken by the US government and governments of other nations to promote internal development and the growth of institutions within those nations. "Institutions" in the context of nation assistance is a broad term, It includes large national agencies such as the HN's military or ministry of public works. It also includes smaller organizations such as local public education and health agencies. The US government conducts nation assistance activities to promote stability within specific countries or regions of the world, as well as for humanitarian reasons.

Nation assistance activities must be a coordinated effort between the HN and all US agencies involved (Department of State, Drug Enforcement Administration, and so forth). Coordination is essential to achieve a synergistic effect rather than isolated accomplishments. Lack of master planning may result in disjointed programs. Examples are bridges with no roads between them, roads without bridges across major gaps, or health facilities with no water supplies. In each HN, the US Ambassador's country team will develop a country plan (the country team concept is described below). The country plan contains a "vision" of what the US and the HN should do together. The HN must fully concur with this vision, as well as the method of achieving it, for nation assistance to be successful.

The HN, in conjunction with the US, must establish objectives for internal development that meet the needs of the nation's people. These objectives may include reducing poverty, distributing the benefits of national development equitably, or taking steps to ensure continued prosperity within a nation. For a developmental assistance program in which the US is participating to be effective, it must meet several criteria. Programs or projects should serve the entire population, not just a small section of the society. Project designs must take into account the skill and level of technology of the recipients. Results must be maintainable (structures or facilities) or sustainable (technologies or capabilities) by the people. A key point here is that a program should attempt to develop an HN's capabilities, rather than simply providing a temporary solution to a problem. For example, planners should plan and execute a bridge construction project in such a way that an HN learns to build its own bridges. The HN should not simply receive a completed project without developing a construction capability.

Expectation management is a key aspect of nation assistance. National development results in rising expectations on the part of the populace. As the people of a nation see the benefits of development accrue to others, they will want more for themselves. While this is understandable, it may not be possible for a government to meet the people's desires on a schedule that satisfies them. The government and the people of the HN must not be given any promises that the US may not be able to fulfill. To do so may cause (or worsen) resentment toward the US.

Nation assistance should not be viewed only as a counterinsurgency strategy. However, failure to develop the type of conditions a nation assistance program promotes may create an environment which favors insurgency. Nation assistance activities should strengthen the image of the national and local governments in the eyes of the populace. This increases the legitimacy of the government, fosters national unity, and promotes stability. Nation assistance programs must be designed so that the perception of the HN people (as well as the actuality) is that their government is the guiding force. Some of the activities that nation assistance encompasses include disaster relief, humanitarian and civic assistance (HCA), combined exercises, deployments for training (DFTs), and some types of security assistance. Some of these nation assistance activities are addressed in this chapter and Chapter 4.

COORDINATION OF US ASSISTANCE ACTIVITIES WITHIN A HOST NATION

US Army engineers operating independently in an HN coordinate their activities through the US country team in that country. The US country team consists of the principal representatives of the US government departments and agencies in an HN (see Figure 2-1). These may include the Department of Defense (DOD), the US Agency for International Development (USAID), the

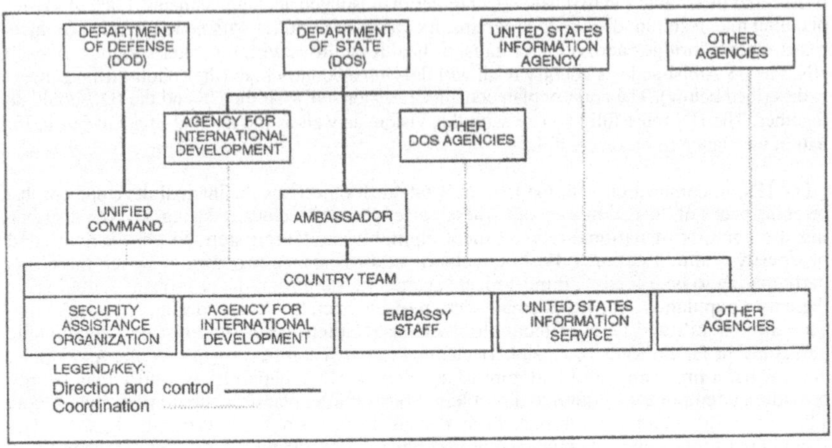

Figure 2-1. Country team organization.

Drug Enforcement Administration (DEA), and so forth. The country team is the executive committee of the embassy. It is headed by the chief of mission (normally an ambassador). The ambassador is the senior representative of the US government in an HN. The country team is the principal organization through which the diplomatic mission conducts operations. It is an informal organization, however. It has no set size, structure, or guidelines for operations. The country team functions as the ambassador's "cabinet." It makes recommendations, but it is not a decision–making body. The manner in which the team operates depends on the desires of the chief of mission. Military representatives on the country team include the defense attache, individual service attaches, and the chief of the security assistance organization (SAO).

The term SAO refers to all US armed-forces organizations that have security assistance responsibilities and are permanently assigned to an overseas US diplomatic mission. The US tailors each SAO to the needs of the HN. For this reason, there is no standard SAO. A large SAO normally has Army, Navy, and Air Force sections, each responsible for accomplishing its service's portion of security assistance activities. A small SAO is divided by function and has no separate service sections. Figure 2–2, page 2-4, shows the typical organizational structures of SAOs.

The SAO in an HN may be known by any of a number of names. The specific name will depend on the number of persons assigned, the SAO's functions, and the desires of the HN. These names include "joint military advisory group," "joint US military group," "US military training mission, " "defense field office," or, "office of defense cooperation." The SAO oversees all foreign–based DOD elements in the HN that have security assistance responsibilities. The SAO chief is responsible to three authorities—

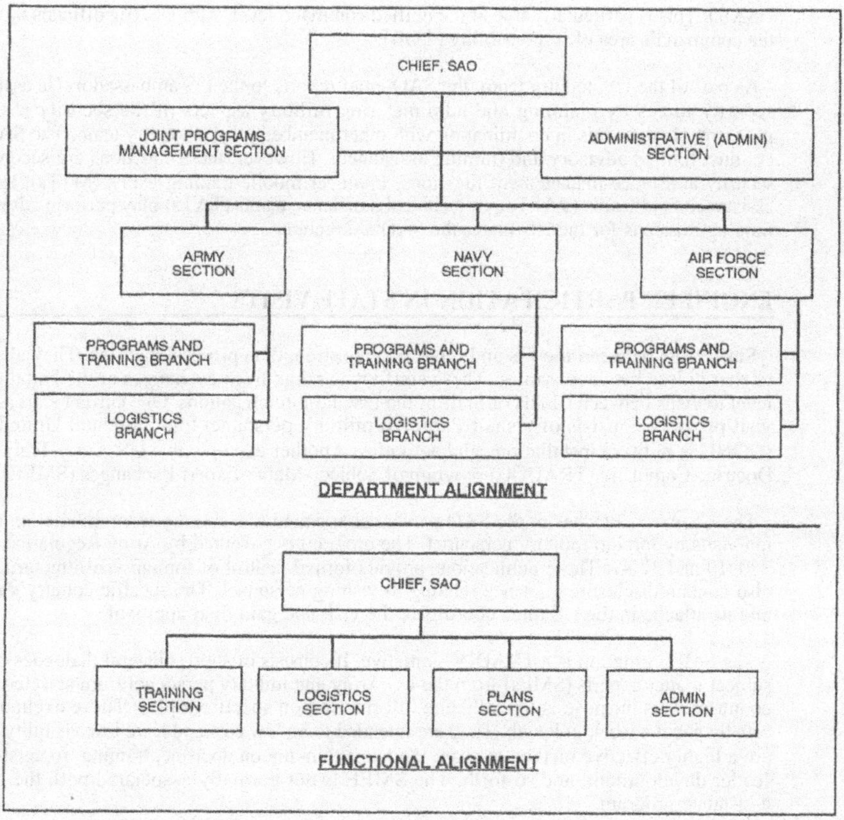

Figure 2-2. Typical SAO structure.

- The Commander in Chief (CINC) of the unified command responsible for the region in which the HN is located.
- The US ambassador in the HN.
- The director of the Defense Security Assistance Agency (DSAA).

Unified commands have a direct role in security assistance activities. The CINC of a unified command integrates all military security assistance plans and activities with regional US military plans. He ensures consistency between the plans. The CINC also provides SAOs with technical assistance and administrative support. He supervises SAO activities in matters that are not functions or responsibilities of the chiefs of US diplomatic missions. He advises the Joint Chiefs of Staff (JCS) on significant activities in his area of operations from both regional and country-specific perspectives. Coordination is essential at all levels between DOD and

USAID. This is particularly true at the unified command level, with USAID officials throughout the command's area of responsibility (AOR).

As part of the US country team, the SAO chief reports to the US ambassador. He assists HN security forces by planning and administering military aspects of the security assistance program. He does this in coordination with other members of the country team. The SAO may conduct limited advisory and training assistance. However, these functions are secondary to security assistance management functions. Engineer mobile training teams (MTTs), technical assistance field teams (TAFTs), or technical assistance teams (TATs) may perform advisory and training missions for the HN under the SAO's direction.

ENGINEER PARTICIPATION IN STAFF VISITS

Staff visits between the US and other nations strengthen professional ties. They also serve to share ideas between armies. These staff visits range from exchanges at the battalion staff level to visits between Chiefs of Staff of the US and foreign nations. One current staff assistance visit program consists of visits by foreign military personnel to continental United States (CONUS) military installations and activities. Another example is a US Army Training and Doctrine Command (TRADOC) program of Subject–Matter Expert Exchanges (SMEE).

The US Army attaches or the SAO personnel assigned to a country team initiate familiariza-ion visits by foreign military personnel. The program is governed by Army Regulations (ARs) 380–10 and 37-47. These publications provide formal control of foreign visit itineraries. They also contain disclosure guidance relating to visiting personnel. The specific country's military and its attache in the US must coordinate the visit and gain final approval.

The SMEE program is a TRADOC initiative. It consists of short, bilateral dialogues between subject-matter experts (SMEs) from the US Army and military personnel from selected foreign countries. Its purpose is to exchange information on specific topics. These exchanges are usually about a week in length. They are intended to be low cost and have low visibility. SMEEs are a highly effective method to share the latest thinking on doctrine, training, force structure, leader development, and so forth. The SMEE is not normally associated with the security assistance program.

Each staff visit is unique in purpose, scope, and level. Most can be funded through the above programs or other similar ones. US military engineers may participate in several ways. They may be members of the staff involved in US and foreign nation staff exchanges. They may be assigned to an installation, activity, or unit that is hosting foreign personnel. In some cases, they may be the SME involved in information exchanges.

ENGINEER PARTICIPATION IN THE US ARMY PERSONNEL EXCHANGE PROGRAM (PEP)

The PEP objective is to foster professional military relationships between the armies of the US and other nations. PEP officers are fully integrated into the HN's army, usually performing the same duties that a member of the host army would. PEP officers fill authorized positions either in the HN or the US. Officers selected for these positions serve two-year tours and must be language and specialty qualified. AR 614-10 provides detailed information on this program.

SECURITY ASSISTANCE TRAINING

The US government uses two methods to provide security assistance training. One method is the International Military Education and Training Program (IMETP), which is funded by the US government. The other method is IMETP–type training that is funded by foreign governments and purchased through foreign military sales (FMS). The objectives of security assistance training are to-

- Develop skills required to operate and maintain US equipment.
- Develop a foreign country's expertise in management of its military.
- Foster development of a foreign country's professional and technical training capabilities.
- Promote US military rapport with counterparts in a foreign country.
- Promote a better understanding of the US, its people, political system, institutions, and way of life.
- Increase the awareness of international military personnel on the US commitment to human rights.

IMETP increases the efficiency, professional performance, and readiness of the armed forces receiving the training. IMETP usually consists of training in CONUS, but training may be conducted in other countries as well. It involves several methods of schooling: formal courses, orientation tours, and on–the-job training. Students are often personnel targeted to assume leadership and management roles in their armed forces and elsewhere in their governments. IMETP is funded yearly by Congress under the overall security assistance program, based on requests by the State Department in consultation with DOD.

FMS allows qualified foreign governments to purchase defense articles, services, and training from the US. The sale of defense articles often requires that personnel be trained on operation and maintenance of the equipment. This is accomplished by two methods-IMETP-type training (separate from IMETP due to funding methods) and mobile training assistance. Mobile training assistance is provided by both MTTs and field training services (FTS). Details for this program are found in AR 12–15.

An MTT is a group of DOD personnel on temporary duty to train foreign personnel. MTTs provide the HN requesting the team with a self–training capability in a particular skill. The team trains selected HN personnel. They then become an instructional base within that country. The team's size and composition are based on the request submitted by the HN. Engineer-specific MTTs may relate to any aspect of combat engineering, sustainment engineering (which includes construction), or topographic engineering. Sometimes the skill may not be purely military in nature, such as training in locating and developing water supplies. The SAO, working in conjunction with the country team and the HN, can determine what specific capabilities the HN requires. MTTs are requested by an HN through the SAO. MTTs are programmed to last 179 days or less, including all the time that members are away from their

home station. The SAO controls the operations and the administration of MTTs while in the foreign country.

An FTS is basically a long–term MTT either using Extended Training Service Specialists (ETSS) or contract field services (CFS). ETSS consists of DOD military or civilian personnel qualified to provide training in installation, operation, and maintenance of FMS equipment. ETSS personnel are assigned (usually limited to one year) to the in-country SAO. CFS provides the same services; however, team members are contracted from private industry. They are only used when DOD personnel are unavailable or their use is impractical.

TECHNICAL ASSISTANCE TEAMS

TATs consist of DOD personnel on temporary duty. They provide in-country technical assistance to the armed forces of a foreign government on specific weapons, equipment, technology, or supporting systems. Assistance is provided for a relatively short time and for a purpose which cannot be satisfied by CONUS training programs or commercial contracts. The principal purpose is technical assistance, not training, although some instruction may be provided to HN technicians. AR 12–7 explains policies, procedures, and restrictions relating to TATs.

ENGINEER ADVISORS

Engineer advisors are individuals who provide assistance to an HN engineer unit. These individuals advise engineer units to enable them to train and operate more effectively. Advisors do not supplement or supercede the chain of command. They are attached to engineer units to provide assistance on combat, sustainment, or topographic engineering depending on unit needs. Individuals in these positions must be able to speak the language of the HN and understand its political, economic, and social conditions.

EXERCISE DEPLOYMENTS

Exercise deployments include a diverse group of peacetime activities conducted by the US Army. Deployments can range from a Special Forces (SF) team to a well-drilling detachment to an exercise involving one or more engineer battalions. An example of this last case is the Fuertes Caminos (Strong Roads) series of exercises. Deployments are funded through the JCS exercise program or by individual services to support US security concerns worldwide. Exercises are planned and directed through the unified commands and involve the Total Army (Active, Army Reserve, and National Guard), other services, and allied forces.

Engineer units may participate in major deployments designed to exercise contingency plans in support of existing alliances. Engineer training activities in these large-scale exercises often involve combat engineer support to maneuver units and may occur in more developed areas of the world. Other deployments may be smaller in nature and provide a more subtle approach to furthering US goals abroad. Many deployments to Third World nations fall into this category.

Deployments for exercises in the Third World environment have unique characteristics. These deployments will include conventional and special operations forces (SOF) of the Army and other services. Engineer units from the US Air Force, US Navy, and US Marine Corps can often provide capabilities not found in Army engineer units. These units from the other services will complement Army capabilities. Nonmilitary agencies including the State Department (such as the ambassador and country team), the Justice Department, the DEA, and others also have prominent roles. Exercises in the Third World often support a government fighting an insurgency or potential insurgency. This US role is defined as foreign internal defense (FID). Specific considerations for FID will be discussed in Chapter 4.

The primary objective of these exercises is training. They will often be joint or combined in nature. Exercise-specific equipment, residual repair parts, and other supplies must return to the US with the participating units. However, other security assistance activities may occur in conjunction with exercises. Activities such as equipment transfers and HCA take advantage of the US military presence to help foreign nations. Engineer equipment purchased through FMS or given to allies from excess equipment stocks can be deployed with US engineer units. HN units can be trained on the equipment, and it can be left to improve HN-engineer capabilities.

Exercise-related construction (ERC) principally supports an exercise but may give an HN a residual benefit. ERC is construction on other than US-owned or -controlled real property of an unspecified minor construction project, mainly in support of an in–progress or planned Chairman, Joint Chiefs of Staff (CJCS) exercise outside the continental US, which results in facilities that remain after the end of that exercise. ERC often includes construction or upgrade of roads, airfields, and well drilling and construction of rudimentary ports and pipelines. All engineer sustainment operations that are a part of these exercises are performed to theater of operations (TO) standards.

HCA provides a way through which US military personnel and assets augment other US nonmilitary programs. It improves the quality of life in an HN through rudimentary construction, health care, and sanitation programs. HCA may involve the use of a substantial amount of materials to complete the project. These materials (building materials, machinery, or other items) result in semipermanent structures used for schools, community buildings, or medical centers. They are also used for wells (for communities with pumping equipment) and medical, dental, and veterinarian services. Materials are funded through FMS or USAID or other monies available under the Foreign Assistance Act (FAA). In 1986, Congress modified Title 10 of the US Code to authorize HCA in conjunction with any US military activity. Before this, it was authorized only during JCS-directed or -coordinated exercises.

Military engineers have a major role in exercises conducted in Third World countries. Support to maneuver units may occur, but often the presence of combat units is not desired due to political sensitivities. Engineers frequently comprise the bulk of units involved. They often form a task force headquarters to control the exercise. (Due to transportation costs and limitations and HN sensitivities, engineer unit deployments may be limited to platoon or company (–) in size.) Combined engineer training activities with theHN military, the civilian authorities, and in some cases the civilian population include the following areas:

- Road construction.
- Damage control (natural disaster or man-made destruction).

- Production of construction materials (includes crushed rock, lumber, asphalt, and concrete).
- Location of potable-water sources (or water that can be purified). These may be surface or underground sources.
- Airfield construction and upgrade.
- Well drilling.
- Provision of diving teams for all types of operations.
- Pipeline design and construction management.
- Power–plant operation and maintenance.
- Forestry operations.
- Port construction (developing, rehabilitating, and maintaining port facilities).
- Training on the construction, operation, and maintenance of all types of utilities.
- Fire-fighting training.
- Topographic engineering (mapping, surveying, and terrain analysis).

Exercise planning and execution must consider many regulatory documents such as—

- Joint Staff publications.
- Service regulations.
- US government agency regulations (Department of State, DEA, Customs, and so forth).
- Department of State in-country rules.
- HN laws and Status of Forces Agreements (SOFAs)
- Security assistance regulatory controls.
- US laws (such as the Foreign Assistance Act of 1961, as amended).

The many regulations and controls make coordination with numerous headquarters and agencies essential. At the unit level, much of the coordination will be done by higher headquarters. However, this does not preclude the need to be aware of all the peculiarities of these exercises.

Engineers planning for exercise deployments to Third World countries must address the following issues:

- Unit mission for the exercise. A unit's mission statement and capabilities should match the types of activities it will execute during the exercise.
- Exercise objectives. All exercise objectives must be clearly defined and stated up front. The unified command's and HN's perceptions that the focus of an engineer deployment is either "nation assistance" or "training" must be resolved to the satisfaction of all. This is essential because the focus of the deployment will affect the approach that US participants will take. The primary goal of a training exercise is to train US soldiers. The primary goal of a nation assistance project is to transfer a skill, as well as an end product, to the HN. While the result of each type of deployment maybe a finished project, the approach will be different, based on the focus.
- Exercise focus. The policial element of national power has primacy in the exercise area. Close coordination is required, either through higher headquarters or directly to the SAO.
- Area of operations. The exercise area may not correspond to the CAPSTONE assignment (wartime mission area) of the engineer unit. Special language and regional orientation training may be required.

- Force security. Regardless of the expected potential for violence, this is of paramount importance to the political element. The chain of command must clearly state the rules of engagement (ROE). The ROE must be understood by all exercise participants.
- Skill requirements. Planners must identify the required levlel of experise in project design, quality assurance, and construction management. The current skill level of unit personnel must be determined and raised (if necessary) before the exercise.
- Equipment requirements. Modification table of organization and equipment (MTOE) equipment may not be adequate for exercise projects. In some cases, additional equipment may be available through temporary loan. If not, commercial equipment must be contracted, as well as operator and maintenance training.
- Logistical requirements. Although these exercises are designed primarily as training events, they are heavily logistics-oriented. This is especially true in the area of contracting.
- Funding issues. Because of the variety of funding methods for exercises, budget and contract issues must be addressed early when planning a deployment. It is essential to use proper funding authorizations. The two major types of funding involved in exercise deployments are FAA funds and operation and maintenance (O&M) finds. Within one deployment, it may be necessary to use more than one source. Funding restrictions are subject to change as legislation and US policy are modified. Commanders must consult their legal staff for current guidance.
- Legal considerations. Claims and HN laws must be addressed in planning and project management.
- Communications. Engineer-unit communications equipment is often inadequate because of dependence on communications assets of higher headquarters. Augmentation must be planned for engineer-controlled deployments.
- Responsibilites. A clear understand must be developed between the US and the HN as to who is responsible for each aspect of the exercise. This includes design of projects, material procurement, construction standards, quality control, and so forth.

US Army Reserve (USAR) and Army National Guard (ARNG) engineer units participate in overseas CJCS exercises under the concept of oversea deployment training (ODT). ODT provides the opportunity to conduct wartime mission-oriented training and planning in an overseas location to units deployed in support of JCS–approved operations plans. It also provides CONUS–based units with the opportunity to conduct mobilization, deployment, reception, execution of training, and redeployment activities at outside continental United States (OCONUS) locations. Another ODT objective is to strengthen CAPSTONE relationships between units. AR 350–9 outlines policies and procedures relating to ODT.

An exercise involving a major deployment represents one of the most challenging training events for military engineers. The opportunity to design, plan, and perform actual construction to TO standards in peacetime is invaluable to engineer units in the active, reserve, and National Guard. Exceptions to the "TO standards" approach to construction will result if the HN plans to use the finished product of a project for more than two years. An initial design for a project may be more robust than the TO standard, or the project designer may develop a maintenance program for HN implementation.

CIVIC ACTION PROJECTS

Engineer-related civic action projects are part of the overall Military Civic Action Program, which is defined as follows: The use of preponderantly indigenous military forces on projects useful to the local population at all levels in such fields as education, training, public works, agriculture, transportation, communications, health, sanitation, and others contributing to economic and social development which would also serve to improve the standing of the military forces with the population (see Joint Publication (Jnt Pub) 1-02).

This definition distinguishes military civic action from civic action conducted by the Department of State (through USAID, for example). The definition also separates military civic action from humanitarian assistance that the US military conducts in response to natural disasters. Military civic action will have a lasting effect only if the local populace feels that it is part of the process. A completed project has greater impact if the people feel that they helped identify the need for the project and participated in completing it. The resultant product of any military civic action project must be maintainable by the HN. Projects that deteriorate through neglect (or the inability of the country to do adequate maintenance) are of no use to the HN.

Military civic action contributes to the development of a nation and improves the standing of its military in the eyes of its people. It helps HN forces mobilize the people and other resources in support of the HN's government. An additional benefit includes improving perceptions of the US government presence and attitudes toward US citizens in general. Military civic action may also serve to transfer US values to the HN military-particularly the concept of civilian control of the military.

Military civic action manifests itself in many forms. The US Army Corps of Engineers (USACE) has a dual role as a military entity and a public–works agency. This combined military and civil engineering potential allows USACE to respond to the needs of the Army and the nation as a whole. In this peaceful international environment, there are ample opportunities to use both capabilities.

The following are examples of US support to military civic action:

- Engineer personnel can assist foreign military engineer units with "self–help" programs. HN personnel may learn about building roads, surveying, equipment operations, water supply and distribution, vertical construction, and other skills.
- USACE can establish field offices that serve as a training ground for indigenous engineer personnel. The HN forms a cadre of experts who can administer developmental programs.
- US Army engineer units can directly constructor develop infrastructures. This often occurs as part of some other activity such as a unit deployment for training or a combined training exercise. Military civic action funds are used to pay for portions of small projects built by US forces as part of these activities

The Civic Action Team (CAT) Program is a unique approach to military civic action. It is a joint service project in the Pacific, specifically in the Freely Associated States (FAS) and the Trust Territory of the Pacific Islands (TTPI). The program is managed by the Commander in Chief, US Pacific Command (USCINCPAC). It has two objectives. The first is to assist and train the local population by undertaking projects that are of particular interest to the government of the FAS/TTPI. The second objective is to maintain a favorable military presence in the region. The projects are executed by detachments that work with the local populace. A detachment generally consists of an engineer officer in charge (OIC), a noncommissioned officer in charge (NCOIC), a medic, and eleven soldiers with carpentry, electrical, plumbing, and equipment-operating skills.

ENGINEER CONSULTING AND CONSTRUCTION MANAGEMENT SERVICES

Engineer consulting and construction management services in a peaceful strategic environment are basically military civic action on a larger scale. They may also be conducted in an area where a uniformed military presence is not practical or desired. International construction projects are included in the FMS program. A foreign country desiring to develop its infrastructure or improve the quality of life for its citizens can purchase construction services from the US government under FMS. Funding is accomplished using a variety of methods ranging from cash payments to US loans and grants. Overall control of this effort is performed by the DSAA.

Construction projects are initiated by the HN, included in the country team's Annual Integrated Assessment for Security Assistance (AIASA), and funded by the HN or by congressional annual appropriations. USACE acts as the construction manager for the US government for these construction projects. The HN gains fringe benefits from the technical advice and assistance provided by USACE. Primary among these are the following:

- Untrained local employees, through association with those trained by USACE personnel, receive otherwise unattainable practical on–the-job training in various skills and trades.
- USACE and contractor personnel enhance the economy through local purchase of material and services.
- A sense of pride is instilled in the population through the development of an in-country construction and professional engineer capability that may have been previously nonexistent.
- Resources, previously ignored, are developed and recognized as valuable assets to the nation's welfare.

For small projects, construction is performed by the HN's military engineers to the greatest extent possible to support the philosophy of military civic action. For projects that exceed the capability of the HN's military, HN or US civilian contractors may be used. Local materials are used whenever possible to ensure familiarity for HN personnel. (Note: A waiver is required when non–US materials are used for the project that is funded by the US.) A comprehensive assessment of the country's military engineer expertise and capabilities and the availability of construction materials is a critical step in the management process. An engineer MTT may be required to increase the indigenous military engineer capabilities. In many areas of the world, other US military services are more readily available to provide this engineer training (for example, Navy or Coast Guard).

When the proposed construction project exceeds indigenous military engineer capability, then USACE contracts a civilian construction agency to do the work. As before, local civilian construction firms are used when possible. High-dollar projects (over $2.5 million) incur congressional oversight.

There are many examples of US support for international construction. Long-term USACE involvement in projects in Saudi Arabia is an excellent example of a long–term, high-dollar–value operation. The Saudi Arabian government purchased the services of USACE as their construction management agent. This included the formation of the Middle East Engineer Division, due to the projects' scope. At the low end of the spectrum (dollar wise) was a $40,000 medical clinic constructed in Mali by indigenous military engineers. In all cases, USACE aids various agencies, both US and HN. It provides construction management expertise, locates and obtains materials, and implements quality control.

TOPOGRAPHIC ENGINEERING SUPPORT

The Defense Mapping Agency (DMA) is the DOD agent responsible for fostering international cooperation and for negotiating and concluding agreements in the mapping, charting, and geodesy field. DMA has contact with mapping organizations in almost 200 countries and has formal agreements with over 70 governments. These agreements can be bilateral or multinational and may involve product exchange, cooperative production programs, technical assistance, standardization, and facsimile reproduction. Field Manual (FM) 5–105 explains responsibilities and definitions for international mapping, charting, and geodesy cooperation. Additionally, it establishes the doctrinal organization, capabilities, and employment of US Army topographic engineer units in a TO.

Chapter 3

Disaster Relief in the US and Its Territories

"During January of this year [1977] snow fell on Buffalo every day leaving total accumulation of more than 60 inches of snow. An average temperature of 13 degrees for the month added to the problems. As the blizzard continued, high winds blew over the city causing low temperatures and enormous snow drifts. Eventually winter's grip on the city left many of its automobiles as well as citizens stranded.

Thomas R. Casey, of the Federal Disaster Assistance Administrtation, was appointed by President Jimmy Carter to coordinate efforts with the state of New York. In close evaluation of the situation, which worsened by the day, Casey called on active Army units to join the [disaster relief] effort. Mr. Casey contacted the First US Army Disaster Control Element and requested assistance. First Army notified the US Army Forces Command (FORSCOM) of the request.

On the morning of January 29, FORSCOM asked XVIII Airborne Corps and Ft. Bragg headquarters for a list of all engineer equipment available to assist in the snow removal operation. Within hours, the 20th Engineer Brigade (Combat) (Airborne) provided the list and placed the equipment and personnel in "ready" status. "

"Operation Snow-Go to the Rescue"
Specialist 4 Robert Small
The Engineer, Jan/Feb/Mar 1977

This chapter addresses conditions under which the military may help civil authorities during emergencies and major disasters. It also outlines policies, procedures, and types of support that may be provided.

An emergency is any occasion or instance that requires federal assistance to supplement state and local efforts and capabilities to save lives; protect property, public health, and safety; or to lessen or avert the threat of a catastrophe. An emergency declaration by the President provides short–term programs for immediate response. These programs provide a wide variety of life-saving, public health, safety, and property–protecting measures. The total assistance provided for any emergency declaration may not exceed $5,000,000, unless it is increased by the Federal Emergency Management Agency (FEMA) Associate Director.

A major disaster is a hurricane, earthquake, fire, explosion, or other catastrophe of a serious nature. The President can declare a major disaster when an event has caused damage severe enough to require federal assistance. This assistance supplements the efforts and available resources of states, local governments, and disaster–relief organizations to alleviate damage, loss, hardships, or suffering.

MILITARY ASSISTANCE TO DISASTER RELIEF

Responsibility for disaster relief lies primarily with individuals, private industry, state and local governments, the American National Red Cross (ANRC), and federal agencies designated by statute. Department of Defense (DOD) components may assist civil authorities, recognized relief agencies, and federal agencies charged with disaster relief. Assistance is provided as needed or as directed by higher authority. Federal military assistance may be provided when—

- A situation exceeds the capabilities of state and local governments.
- Support is not available from commercial sources. Military support will not normally be provided if it competes with private enterprise or the civilian labor force, unless directed by FEMA.
- Requesting authorities make a commitment to reimburse the federal government.

In some cases, a serious emergency or disaster is so imminent that waiting for instruction from higher authority would preclude an effective response. A military commander may take action to save human lives, prevent suffering, or lessen major property damage. The commander will report the action to a higher authority as soon as possible. He will request guidance if continued support is necessary or beyond his ability to sustain. Under these conditions, the commander will not delay or deny support pending receipt of a reimbursement commitment from the requester. Army Regulation (AR) 500-60 contains guidance for these situations.

Military resources used in civil disaster relief are limited to these not immediately required for military missions. Active federal forces used during disaster relief remain under the control of their military chain of command. Civil agencies exercise no command authority over them. Military support will consist of the minimum required assets. Support will end as soon as possible. Military assistance for rehabilitation after a disaster occurs only when directed by FEMA or in support of operations conducted by United States (US) Army Corps of Engineers (USACE).

AUTHORITY AND RESPONSIBILITIES

The following laws and regulations establish statutory authority and limitations for disaster-relief activities:

- Public Law (PL) 93–288, as amended (The Stafford Act)(42 US Code 5121 et seq). Authorizes FEMA to direct DOD to use available personnel, supplies, facilities, and other resources to provide assistance after a major disaster or emergency declaration by the President. Immediately after an incident that may qualify as a major disaster or emergency, FEMA can direct DOD to use personnel and equipment to remove debris and wreckage and temporarily restore essential public facilities and services. This is permitted when there are threats to life and property that cannot be dealt with effectively by state or local governments. The performance of the emergency work may not exceed a 10-day period from the date of the mission assignment. Under The Stafford Act and The Federal
- Response Plan, USACE has a standing mission assignment for the functional area of Public Works and Engineering (Emergency Support Function #3) in response to a catastrophic

earthquake or other major disasters. In addition to this mission, USACE supports eight other emergency support functions.

- Public Law 84-99 (Flood and Coastal Storm Emergencies) (33 US Code 701n). Authorizes USACE to provide emergency and disaster assistance. It authorizes the Chief of Engineers, acting for the Secretary of the Army, to undertake activities including disaster preparedness, advance measures, and emergency operations (flood response and postflood response). It also authorizes rehabilitation of flood-control works threatened or destroyed by a flood, protection or repair of federally authorized shore-protective works threatened or damaged by a coastal storms, provision of emergency water due to a drought or contaminated source, and emergency dredging.
- Section 686, title 31, US Code (The Economy Act). Authorizes all executive departments or independent government establishments to order materials, supplies, equipment, and work or services from each other. It does not specifically address disaster relief, but federal agencies can use it to obtain military support. It can provide military assistance to federal agencies in situations not otherwise provided for by law. The Economy Act transactions must be coordinated through the Office of the Special Assistant to the Secretary and Deputy Secretary of Defense.
- Section 3, Title 36, US Code. Authorizes the ANRC to provide disaster relief to persons suffering from fires, floods, and other natural disasters.
- Army Regulation 500-60 (Disaster Relief). Prescribes Department of the Army (DA) disaster–relief policies and procedures and assigns responsibilities for such activities.
- Engineer Regulation (ER) 500-1-1 (Natural Disaster Procedures). Prescribes policies, guidance, and procedures for the USACE Domestic Emergency Program (Disaster Preparedness and Emergency/Disaster Response) under the authorities of PL 84—99, The Stafford Act (as amended), and AR 500-60.

Within the DOD, the following individuals have disaster-relief responsibilities:

- Secretary of the Army. The Secretary of the Army is the DOD Executive Agent for military support in presidentially declared major disasters and emergencies in the US. He is responsible for the use, coordination, and control of military resources used by DOD components.
- Commander, FORSCOM. The Secretary of the Army delegated to the Commanding General (CG), FORSCOM, the authority to task DOD components for resources to conduct disaster relief in the continental United States (CONUS). The CG, FORSCOM, can further delegate this authority to CONUS Army commanders, but no lower.
- Commanders of Unified Commands (in Alaska, Hawaii, and US Territories). The Secretary of the Army delegated the authority to conduct disaster-relief operations outside CONUS to commanders of unified commands. They are responsible for disaster–relief operations in their geographical areas of responsibility.
- Commanders of CONUS Armies. As specified in AR 500-60, CONUS Army commanders will—

  –Plan for and conduct disaster–relief operations in their areas of responsibility.
  –Appoint a DOD military representative (colonel or above) as disaster control officer (DCO). The DCO will be the single point of contact for the Federal Coordinating Officer (FCO) during each declared emergency or disaster.

- Control DOD resources furnished for disaster relief. Establish and maintain disaster–relief liaison with appropriate federal, state, and local authorities, agencies, and organizations.
- Furnish resources (on request) to USACE major subordinate command (MSC) and district command (DC) engineers conducting a flood fight.
- Furnish resources (on request) to support a USACE response to FEMA assignments.

The CG, FORSCOM, has established an emergency operations center (EOC) in each of the CONUS Armies. On notification of a pending or declared disaster, the CG, FORSCOM, tasks the appropriate CONUS Army to activate its EOC.

- Commanding General, USACE. Ensures that USACE MSC and DC engineers provide disaster–relief assistance when required by an imminent serious condition, as required by statute, or as directed by FEMA under PL 93–288. Specifically, the CG, USACE, will—

  - Ensure that USACE MSC and DC engineers establish and maintain liaison with the CONUS Army or unified commanders, FEMA regional directors, United States Coast Guard (USCG), ANRC, and other federal, state, and local government agencies.
  - Furnish commanders with information on floods or other disasters or emergencies. Information on USACE activities should be included.
  - Ensure that USACE procedures for disaster relief are coordinated with the CONUS Army or unified commanders.
  - Provide a DCO, as required by the CONUS Army or unified commander, to coordinate DOD relief effort during an emergency or major disaster. The DCO will establish priorities and define work areas for DOD elements and USACE contractors in coordination with the FEMA regional director.
  - Ensure development of procedures for engineering, construction management, and contracting support to troop units involved in disaster–relief operations, when requested.

DOD personnel will conduct disaster-relief operations in conjunction with personnel from state and other federal agencies. The following agencies and individuals are responsible for specific disaster–relief activities:

- Federal Emergency Management Agency. FEMA, under Executive Order 12148, administers The Stafford Act and prescribes rules and regulations for carrying out the provisions of the act. The FEMA associate director or regional director may direct any federal agency to provide emergency assistance in save lives and to protect property, public health, and safety. Assistance may be provided with or without compensation.
- Environmental Protection Agency (EPA). Coordinates the response by federal departments; federal, state, and local agencies; and private parties to certain environmental threats. EPA coordinates efforts to control oil spills or the threats of oil spills into navigable waters. It also coordinates efforts to control hazardous substances that present a threat to the environment.
- Department of the Interior. Protects lands under its administaration from fire damage. This responsibility is assigned specifically to the Bureau of Land Management, the National Park Service, the Bureau of Indian Affairs, and the US Fish and Wildlife Service.
- United States Coast Guard. Protects lives and property in maritime disasters. It also provides an on-scene coordinator (OSC). The OSC coordinates the response of federal,

state, and local agencies and private individuals to control oil spills or other hazardous substances in coastal or contiguous waters.

- Forest Service, Department of Agriculture. Protects forests and watershed lands from fire, in cooperation with other agencies.
- Boise Interagency Fire Center (BIFC). BIFC is a federal center under the control of the Departments of Agriculture and the Interior. It provides national coordination and logistical support for federal fire control. DOD provides emergency assistance to the BIFC to help suppress wildfires within CONUS. AR 500-60 contains a copy of the memorandum of understanding between DOD and BIFC.
- Public Health Service (PHS). Uses its health resources and technical personnel to control communicable diseases. PHS assists federal, state, and local agencies in maintaining safe food supply, sanitary waste disposal, refuse disposal facilities, and the control of insects and rodents. It can also evaluate health hazards and recommend corrective action.
- National Weather Service (NWS), National Oceanographic and Atmospheric Administration (NOAA). Provides storm warnings (tornado, hurricane, heavy rainfall, and so forth), weather forecasts, and flood warnings. It also provides forecasts of water stages, particularly the peak flow and duration of high water.
- National Commuications System (NCS). Provides technical communications support to FEMA in predisaster planning and during presidentially declared emergencies or disasters.
- American National Red Cross. Prevents or mitigates suffering caused by fires, floods, and other natural disasters. In emergency situations where FEMA is not involved, the ANRC may request Army assistance when local resources are inadequate. Local ANRC chapters will normally request Army support through the ANRC area office to the CONUS Army commander. During an imminent serious situation, the local ANRC chapter may request help from the nearest military installation.
- State Emergency Management Agencies. Direct and coordinate state emergency or major disaster–relief activities.

ARMY SOURCES OF SUPPORT

DA may provide assistance for disaster relief through a variety of means. This may include supplies, equipment, or support from selected individuals. At the other end of the spectrum, entire units may conduct disaster-relief operations. These units may come from the Active Army or US Army Reserve (USAR). Under some circumstances, Army National Guard (ARNG) units may be placed in a federal active duty (FAD) status. USACE is a major relief asset for the Army during a disaster.

Active Army units will support disaster-relief operations at the direction of their higher command. Exceptions occur when a serious emergency or disaster is imminent and waiting for instruction from higher authority is inappropriate. Criteria are outlined in the section on military assistance to disaster relief. In these cases, a military commander may do what is required to save human lives, prevent suffering, or lessen major property damage.

USAR units or individuals may perform disaster-relief operations under any of the following conditions (see AR 500–60 for additional guidance and limitations):

- When ordered to active duty after the President has declared a national emergency.
- When ordered by the DA on the recommendation of the CONUS Army commander and the CG, FORSCOM, as annual training.
- When approved by the CG, FORSCOM, in a voluntary active duty for training (ADT) status.
- When USAR commanders approve voluntary participation, with orders, during imminent serious conditions in a nondrill, nonpay status. USAR members taking part in such support are performing an official duty.

ARNG units may perform disaster relief while on federal active duty or state active duty (SAD). When not in active federal service, ARNG units will remain under the control of the state governor. Normally, they will be assigned a mission through their chain of command. However, with the concurrence of the governor, they may accept missions from the CONUS Army or unified commander on a reimbursable basis. Units may take federally owned equipment with them when ordered into disaster–relief areas. USAR equipment may be loaned to ARNG forces for disaster relief while on state active duty status. Procedures are outlined in AR 500-60. Once federalized by the President, units will be under the control of the chain of command established for the specific disaster-relief operation.

USACE responsibilities include-

- Establishing policies and procedures to implement USACE authority to provide emergency and disaster assistance under PL 84-99 and AR 500-60 and in support of other agency authorities.
- Establishing and maintaining an emergency management organization.
- Providing guidance to subordinate elements supporting other agencies with emergency and disaster assistance.
- Maintaining all USACE elements in a state of readiness to respond to disasters and emergencies.
- Maintaining liaison and/or coordinating USACE programs with appropriate federal agencies to ensure timely support to requests for assistance.

Within the Corps of Engineers, MSC and DC commanders have been delegated the overall responsibility for preparedness and response missions.

The MSC commander's responsibilities include-

- Establishing and maintaining an emergency–management organization. This includes appropriate personnel, space, and facilities to manage required preparedness and response programs.
- Developing guidance and procedures to implement emergency– and disaster–response authorities within the MSC through publication of supplements or other documents.
- Establishing and maintaining contacts with appropriate military, federal, and state, agencies and coordinating the USACE emergency program, as necessary.
- Managing the disaster–preparedness program and response activities command wide within delegated authorities.

- Providing guidance to subordinate elements supporting other agencies with disaster assistance.
- Establishing and maintaining inoperable EOC.

The DC commander's responsibilities include-

- Establishing and maintaining an emergency-management organization. It must include appropriate personnel, space, and facilities to manage required preparedness and response programs.
- Establishing and maintaining operational plans and procedures to respond to emergencies and disasters within their authority and geographic area of responsibility.
- Maintaining sufficient, required emergency supplies and equipment.
- Ensuring personnel are prepared to respond to emergencies and disasters.
- Establishing and maintaining an operable EOC.
- Conducting exercises.
- Establishing and maintaining contacts with appropriate officials from military, federal, state, local, and charitable agencies.

DISASTER-PREPAREDNESS PLANNING

Disaster preparedness consists of activities that permit rapid response to natural disasters or emergencies. USACE, FORSCOM, CONUS Army, and installation staffs conduct disaster–preparedness planning to minimize delays.

Within USACE, the Chief of Emergency Management Division or Branch is responsible for USACE MSC and DC disaster–preparedness programs. The following areas are included in USACE disaster preparedness planning:

- Preparing emergency publications. These include regulation supplements, emergency plans, and so forth.
- Conducting disaster–preparedness exercises and training.
- Inspecting nonfederal flood-control works (levees and dams).
- Procuring supplies and equipment for emergency operations.
- Establishing MSC and DC EOCs.

USACE MSC and DC emergency managers develop and coordinate disaster–preparedness programs. Detailed procedures are in Engineer Regulation (ER) 500–1–1.

EMERGENCY OPERATIONS

Requests for Army support for disaster-relief operations may be initiated through several channels including—

- A direct request from a civil authority to a military commander.
- A request from USACE MSC or DC.
- An awareness by USACE or DOD of an ongoing or imminent situation.

- A request from FEMA.
- A request from ANRC.

Military commanders may approve direct requests from civil authorities for emergency assistance to save human lives, prevent suffering, or lessen great destruction or damage. This is outlined in the paragraphs on military assistance to disaster relief beginning on page 3–2. When a commander receives a request that does not meet the criteria in AR 500-60, he should direct civil authorities to forward it through channels to DOD. Requests must contain (see AR 500–60)—

- A brief statement of the situation.
- Assistance being provided by state and local governments.
- Support required and the estimated time needed.
- A statement that military assistance will not be in competition with commercial sources.
- An estimate of costs and provision for reimbursement.
- A statement on whether or not assistance has been requested from another military department or federal agency.

Upon the President's declaration of a major disaster or emergency, the FEMA director may coordinate and administer the support of disaster-relief activities of all federal agencies. FEMA may task federal agencies (including DOD) to provide personnel, equipment, supplies, facilities, and other resources in support of state and local disaster assistance efforts. The Directorate of Military Support (DOMS) is the DOD point of contact in all matters related to military assistance during major disasters or emergencies. DOMS is in the Office of the Deputy Chief of Staff for Military Operations and Plans (ODCSOPS).

When a presidential declaration of a major disaster or emergency is made, FEMA will appoint an FCO. The FCO is responsible for coordinating federal assistance in an emergency or major disaster. He initiates action immediately to assure that federal assistance is provided based on the declaration.

FEMA will notify DOMS or the Army Operations Center (AOC) when a major disaster or emergency has officially been declared. DOMS or the AOC will notify FORSCOM and provide updates as the situation develops. Upon initial notification, FORSCOM will notify subordinate commands and designate a lead and support CONUS Army headquarters. The lead CONUS Army commander will appoint and maintain control of the DCO and staff. The DCO and his staff will deploy to the disaster area and collocate with the FCO.

FEMA regional directors and the FCO will submit requests for DOD resources to the CONUS Army or unified commander through the DCO. Requests for other–than–Army resources will be sent by the CONUS Army or the unified commander to the service headquarters having control of the assets. If USACE representatives receive direct requests while active military assistance is being given, the USACE MSC or DC engineer will notify the CONUS Army or the unified commander. This will be done to ensure coordinated military effort.

Under the Federal Disaster Plan, USACE is the primary and/or lead agency for Emergency Support Function #3, Public Works and Engineering. USACE will provide immediate supplemental engineering and construction support to the state following a presidential declaration. This includes water supply. Some typical response activities include-

- Emergency debris removal.
- Temporary repair or replacement of roads and bridges.
- Emergency restoration of critical facilities and services.
- Demolition or stabilization of damaged structures.

In addition to its mission under the Federal Disaster Response Plan, USACE is also one of the federal agencies FEMA tasks to provide engineering, design, construction, and construction-contract management in support of recovery operations. FEMA may request USACE to perform the following:

- Direct Federal Assistance (DFA). DFA is the performance of eligible work when a state or local government lacks the capability to perform or contract for the work or the regional director determines that assistance is necessary to meet an immediate threat to life, health, or safety.
- Preliminary Damage Assessment (PDA) and/or Damage Survey Report (DSR). A PDA provides a categorized description and initial cost estimate of damages. FEMA uses the PDA in evaluating a request for federal assistance and to identify major work categories. A DSR provides a specific statement of the extent of damages and a complete description of the damaged facility. It includes a statement of the scope of eligible work and a detailed estimate of the reasonable cost of eligible work. This is the primary vehicle used by FEMA to fund or reimburse applicants for public assistance under The Stafford Act.
- Technical Assistance. Technical assistance is the provision of engineering support, construction management, and technical advice to state and local agencies.

Under PL 84-99, USACE provides flood emergency or disaster assistance in the form of advance measures, flood response, and postflood response.

- Advance Measures. USACE may perform activities to protect against loss of life and damages to either urban areas or public facilities due to flooding. Assistance to prevent damages may be taken before a riverine flood, a hurricane or coastal storm, a storm on an inland body of water, or a closed-basin flood. An imminent threat of unusual flooding must exist to justify USACE advance measures. Advance-measures assistance may be technical or direct assistance. Technical assistance consists of providing any combination of technical review, advice, or recommendations to state and local agencies before, during, or after a flood event. Direct assistance authorizes USACE to supplement state and local resources with its supplies and equipment. As an alternative, USACE may contract construction of temporary flood-control projects.
- Flood Response. Technical assistance consists of providing review and recommendations in support of state and local flood-fighting efforts. This includes addressing flood-fighting techniques and emergency construction methods; inspecting flood-protection structures; and providing hydraulic, hydrologic, and geotechnical data and advice. Direct assistance may include supply issuance, equipment loan, rescue operations, direct flood-fighting operations, and emergency contracting.

- Postflood Response. For USACE to provide emergancy assistance, a written request from a state governor to the DC commander will be provided concurrently with or immediately after the governor's request to FEMA for a PDA. Assistance will be temporary to meet the immediate threat. USACE assistance will be limited to major floods or coastal storm disasters resulting in life-threatening situations. USACE will not take action on a governor's request if it is received subsequent to a Presidential declaration or denial. Assistance may only be provided for a maximum of 10 days from the receipt of the governor's request for assistance. Subsequent requests for additional assistance resulting from the same flood or coastal-storm event will not extend the 10-day period or trigger a new 10-day period. No work will be performed, including contract work, after the 10-day period expires. Assistance may be technical or direct. Activities may include clearance of drainage structures and debris to open critical transportation routes. USACE may also restore public services or facilities and provide other assistance to prevent loss of life or facilities.

Whenever FEMA tasks USACE to perform engineering or construction missions, the USACE MSC commander will notify the CONUS EOC or DCO as appropriate. The MSC and/or DC commander will then determine the capabilities of contractor resources to accomplish the mission. If contractor resources are inadequate, the MSC commander will notify the FEMA regional director and Commander, USACE. He will detail the need for augmentation by regular Army forces. He will also address the impact on completion of the requested mission if assistance is not provided. When both troops and civilian contract elements are involved in construction operations, the designated USACE representative will establish priorities and define work areas in coordination with the FCO. USACE may also request military assistance to respond to an emergency flood event under PL 84-99.

When requested, USACE will provide engineering, construction management, and contracting support to troop units involved in disaster-relief operations. Funding, recounting, and reimbursement for Army participation in domestic disaster–relief operations are explained in AR 500-60.

MILITARY ENGINEER UNIT DISASTER-RELIEF MISSIONS

Military engineers may become involved in disaster-relief operations as individuals, teams, or complete units. Individuals may provide technical assistance to civil authorities in the areas of damage assessment and engineer work estimation. Teams may provide specialized support such as well drilling, power supply and distribution, or utilities repair or reconstruction. Units with personnel and vehicles that perform general tasks may be directed to support all types of disaster–relief operations. (Military personnel will not be used to enforce civil law.) Missions that may require engineer-specific skills and equipment include the following:

- Flood Relief.

  - Assessing damage to roads, bridges, structures, utilities, and so forth.
  - Supporting search and rescue operations with personnel and equipment.
  - Conducting topographic surveys for the extent of flood damage.
  - Overprinting of maps to depict damage, water levels, key facilities, search–and–rescue activities, and so forth.
  - Opening roadways for emergency and medical traffic.

- Constructing temporary bridges.
- Providing emergency power.
- Clearing debris, mud, and so forth.
- Restoring critical facilities, services, and utilities.
- Demolishing unsafe structures.
- Providing and/or restoring power to critical facilities.
- Providing expedient repair of critical distribution systems.

● Earthquake relief.

- Assessing damage to roads, bridges, structures, utilities, and so forth.
- Supporting search–and–rescue operations with personnel and equipment.
- Opening roadways for emergency and medical traffic.
- Restoring critical facilities, services, and utilities.
- Demolishing unsafe structures.
- Providing and/or restoring power to critical facilities.
- Providing expedient repair of critical distribution systems.

● Tornado or hurricane relief.

- Assessing damage to structures, utilities, and so forth.
- Supporting search–and–rescue operations with personnel and equipment.
- Clearing rubbled areas.
- Opening roadways for emergency and medical traffic.
- Restoring critical facilities, services, and utilities.
- Demolishing unsafe structures.
- Providing and/or restoring power to critical facilities.
- Providing expedient repair of critical distribution systems.

● Volcano relief.

- Evacuating threatened areas.
- Assessing damage to roads, bridges, structures, utilities, and so forth.
- Supporting search–and–rescue operations with personnel and equipment.
- Opening roadways for emergency and medical traffic.
- Restoring critical facilities, services, and utilities.
- Demolishing unsafe structures.
- Providing and/or restoring power to critical facilities.
- Providing expedient repair of critical distribution systems.

● Tidal-wave relief.

- Assessing damage to roads, bridges, structures, utilities, and so forth.
- Supporting search–and–rescue operations with personnel and equipment.
- Conducting topographic surveys for the extent of damage.
- Overprinting of maps to depict damage, key facilities, search–and–rescue activities, and so forth.
- Opening roadways for emergency and medical traffic.
- Constructing temporary bridges.
- Providing emergency power.
- Clearing debris, mud, and so forth.
- Restoring critical facilities, services, and utilities.
- Demolishing unsafe structures.

- Providing and/or restoring power to critical facilities.
- Providing expedient repair of critical distribution systems.

- Emergency snow removal.

  - Clearing snow from critical roads and facilities.
  - Clearing access to critical facilities such as hospitals.
  - Hauling and dumping snow away from congested areas.
  - Transporting and delivering critical supplies.

- Forest-fire fighting.

  - Cutting fire breaks with equipment, chain saws, and hand tools.
  - Constructing base camps.
  - Constructing fire roads.
  - Transporting water.

Engineer units supporting disaster–relief operations must take adequate tools and equipment to support all potential missions. Planners must design units' support packages (fuel, water, communications assets, maintenance support, and so forth) to facilitate the success of their operations. A detailed reconnaissance of the area of responsibility is essential in developing a plan of action in conjunction with civil authorities, other federal agencies, and USACE representatives.

When conducting disaster–relief operations, maintaining personnel accountability is a critical task. Leaders must keep track of their subordinates at all times. During a crisis situation, a soldier may become separated from his unit and trapped in a hazardous area. Unit leaders must develop a system for tracking their soldiers at all times (to include during rest or sleep breaks).

RECOVERY AND REHABILITATION

Rehabilitation after a natural disaster will be conducted primarily by civilian agencies. Military assistance is authorized only when directed by FEMA or in support of emergency operations conducted by USACE. FL 84-99 authorizes the Chief of Engineers to develop standards and criteria and to rehabilitate flood-control works or federally authorized and constructed shore-protection projects threatened or damaged by a flood, hurricane, or coastal storm. USACE MSC commanders are authorized to approve these rehabilitation projects. When a project is approved by the MSC commander, information copies will be provided to the Commander, USACE. MSC or DC commanders may disapprove ineligible projects without consulting higher authority. ER 500–1–1 provides specific guidance relating to flood-control work rehabilitation–project procedures.

USACE SUPPORT FOR SPECIAL REQUIREMENTS

PL 84-99 allows USACE to provide water to any locality that is confronted with a contaminated water source. The law authorizes assistance in situations causing or likely to cause a substantial threat to the public health and welfare of the local inhabitants. USACE assistance may only supplement state, local, or combined efforts to provide water for public health and welfare.

Assistance may be provided to localities whose publicly or privately owned water and/or distribution system services a community. Aid will normally be temporary to meet the immediate threat. Permanent restoration of safe water supplies is the responsibility of local interests. USACE assistance will not be used to accomplish deferred or deficient maintenance. Assistance is limited to 30 days or until FEMA begins providing emergency water under its own authorities, whichever is earlier. Headquarters (HQ), USACE may extend this time limit provided there is adequate justification.

USACE commanders will determine water quantity and distribution methods after considering the needs of the individual situation and the cost effectiveness of providing various water quantities. USACE assistance under this authority may include water transport to local water points, distribution of bottled water, and temporary connection of a new supply to the existing distribution system. Installation of temporary filtration equipment or the use of military mobile purification units is also permissible.

During drought conditions, emergency water assistance (usually temporary) is provided to meet minimum public health and welfare requirements. Reviews of requests for assistance are to be tempered by the fact that USACE assistance supplements state and local efforts. Long-term solutions to water supply problems are the responsibility of state and local interests. USACE may construct wells or transport water to farmers, ranchers, and political subdivisions within drought- distressed areas. The Assistant Secretary of the Army for Civil Works determines whether an area is drought distressed. This is based on an inadequate water supply that is causing or is likely to cause a substantial threat to the health and welfare of local inhabitants. This includes the threat of damage or loss of property. Written requests for assistance should come through the governor or his authorized representative.

Water may be transported for human and livestock consumption only. Water for irrigation, recreation, or commercial and/or industrial processing will not be provided under this authority. When USACE assists in transporting needed water, all reasonable transport methods should be addressed. These may include trucks carrying bottled water, trucks or trailers carrying bulk water, or small pipelines laid on the ground. The purchase or acquisition of water, the loading and unloading of water from the transportation mechanism, and the storage facility at the terminal point are not USACE expenses. When the recommended option for transporting water includes a permanent facility, there must be clear justification.

When drilling wells, the project owner will be responsible for providing permanent storage and distribution systems. In situations where water will be provided for livestock, consideration should be given to sell part of the herd to provide resources to satisfy the need.

Any USACE well construction will be paid for by the applicant. USACE may construct wells only when commercial or other sources cannot construct it within a reasonable time. Federally owned well-drilling equipment can only be used when commercial firms cannot provide comparable service soon enough. This is based on the time available while preventing the applicant from suffering increased hardships from inadequate water supply.

Flood hazard mitigation decreases the impact of flooding on people and property by reducing its cause, occurrence, and effects. The primary objective of hazard mitigation is to reduce or avoid federal expenditures resulting from flood situations. This is achieved through a coordinated interagency and intergovernmental team approach in the immediate postdisaster period to quickly develop flood hazard mitigation recommendations. Details of flood hazard mitigation operations are outlined in ER 500–1–1.

The Chief of Engineers is responsible for managing and executing the Army Prime Power Program. He has delegated this responsibility to the US Army Engineering and Housing Support Center (USAEHSC), which is a field operating agency of USACE. The objectives of the Army Prime Power Program are to–

- Provide an insurance level of prime, utility–grade power–generation equipment and local power-distribution capability for theater commanders during military contingency operations.
- Provide advice and technical assistance in all aspects of electrical systems and general power in support of war–fighting, disaster-relief, and nation assistance operations.
- Maintain a cadre of skilled military professionals trained to install, test, inspect, operate, maintain, and support prime power assets.
- Loan prime power assets to high–priority, DOD electrical-power requirements, including disaster relief.
- Provide special support for other federal agencies, such as the Department of State.

Prime power personnel (military occupational specialty (MOS) 52E) are skilled in both military and nonmilitary systems. They can inspect, repair, operate, and maintain nonmilitary prime power equipment, such as commercial diesel plants and distribution systems. This capability is well suited for disaster relief or postcombat operations if power plants or distribution systems have been damaged. USAEHSC can provide generator sets or prime power plants based on the requirements of the requestor. If a requestor is unsure of his requirements, USAEHSC can assist in determining them. This includes load estimates, power systems design, and so forth. USAEHSC has a limited number of soldiers to operate and maintain the equipment or to provide operator training. AR 700–128 provides details on this program.

Under the National Contingency Plan (NCP), DOD is a member of the National Response Team (NRT). Within this organization, DOD is responsible for providing resources for the effective operation of the NCP. The objective of the NCP is to support the National Oil and Hazardous Substances Pollution Contingency Plan. This plan ensures a coordinated federal response at the scene of an oil spill or a release of hazardous substance that poses a threat to public health and welfare. USACE is included in the NCP as a branch of the DOD that has "relevant expertise." It can be used in response to discharges or releases and provides members to the Regional Response Teams (RRTs) according to AR 200–1. USACE support capabilities in cleanup activities include recovery of oil using USACE hopper dredges and the reserve fleet of contractor-owned hopper dredges. USACE can also provide contracting, construction management, real estate support services, engineering, environmental review and monitoring, research and development, tracking and forecasting of oil spill location, and power generation.

USACE participation will normally be at the request of the OSC. However, assistance may be requested by DOD or the RRT chairman. The OSC directs the response efforts at an oil spill or hazardous-substance release. In general, the OSC is predesignated by the regional or district head of the lead agency. However, for releases from vessels or facilities under the jurisdiction, custody, or control of DOD, Department of Energy (DOE), or another federal agency, the OSC/RPM will be provided by the federal agency responsible for the release except for the special conditions stated in the NCP. The USCG will provide an OSC for oil spills, including those from a federal agency facility or vessel within or threatening the coastal zone.

Chapter 4

Support for Insurgency and Counterinsurgency

US Army Engineers made rapid progress paving main route QL13 from Saigon north through III Corps to An Loc. The paving eased military resupply, eliminated casual mining of the road, and assured continuous usability during the rainy season. But the most dramatic change was in the activity of the rural population as the asphalt moved forward. An explosion of commerce was evident everywhere. Vegetables and pigs went south to the cities, while pots, pans, and yard goods went north to the once nearly isolated villages, first by animal carts, then Lambretta scooters, Citroen buses, and GMC trucks. The Viet Cong threat to villages was broken by the increased responsiveness of Vietnamese government forces and services over the improved road plus the realization by the population of an improving quality of life through the lively increase in commerce.

Operations Order 5-69
1st Engineer Battalion, 1st Infantry Division
Lai Khe, South Vietnam, 13 November 1969

United States (US) Army engineer support for insurgency or counterinsurgency operations may range from an advisory role to support to ground forces. These may be both conventional and special operations forces (SOF). The type of support provided depends on several factors:

- Type of operation (insurgency or counterinsurgency).
- Degree of US involvement.
- Needs of the group being supported.
- Capabilities of the group being supported.
- Capabilities of opposing forces.
- Secrecy requirements. Operations may be overt, covert, or clandestine.

All engineers must understand the goals of an operation in which they are participating. This applies not only to military goals but also the political, economic, and informational ones as well. Engineers cannot operate in a vacuum. They must be fully integrated into a well-coordinated master plan. This plan will normally be developed by one of the other agencies of the federal government. The Department of Defense (DOD) will be in a supporting role.

INSURGENCY OPERATIONS

The US may support selected insurgencies that oppose oppressive regimes. The US coordinates this support with its friends and allies. Because support for an insurgency is often covert, many of the operations connected with it are special activities. Due to their extensive unconventional warfare (UW) training, Special Forces (SF) are well suited to provide this support. General–purpose forces may assist when the situation requires their functional specialties.

To be successful, insurgencies rely on the mobilization of personnel and resources from within a country. An insurgent organization must build its legitimacy. Therefore, their efforts must also include political, social and, when possible, economic development, especially in areas under insurgent control.

When US forces are directed to do so, they will provide equipment, training, and services to insurgent forces. The following are types of operations in which US forces can assist insurgents:

- Intelligence gathering.
- Surreptitious insertion.
- Sabotage.
- Subversion.
- Linkup.
- Evasion and escape.
- Institutional and infrastructure development.
- Psychological operations.
- Resupply operations.
- Recruitment, organization, training, and equipping a force to perform guerrilla warfare.

During support to an insurgency, SF units will primarily use their organic engineer personnel. Soldiers provide engineer-related advice, training, and assistance to the insurgent forces. Field Manual (FM) 31–20 and branch-specific manuals provide doctrine relating to these personnel.

Conventional engineer units may support SF involved in these operations. They will primarily assist from locations outside the SF area of operations. SF support bases, located in nearby countries, may require sustainment engineering support. Conventional engineer forces may provide specialized training to either SF personnel or insurgents. Topographic engineer support may be provided in the form of standard or special products (maps, terrain analysis products, and so forth). Vertical and horizontal construction by conventional engineer units may play a key role in these operations.

COUNTERINSURGENCY OPERATIONS

This section describes counterinsurgency operations. These are military, paramilitary, political, economic, psychological, and civic actions taken by a government to defeat an insurgency. The measure a government takes to free and protect its society from subversion, lawlessness, and insurgency are known as internal defense. The action a government takes to promote its growth by building viable institutions (political, economic, social, and military) to meet its needs is known as internal development. Together these activities are known as a nation's internal defense and development (IDAD) strategy. This strategy has two goals. The first is to correct, through internal development, conditions that cause hardship within a nation's society. These conditions may create an environment in which an insurgency can develop. The second goal is to develop an internal defense capability. This also provides the ability to defeat an insurgency should one develop. IDAD is ideally a preemptive strategy-every effort will be made to promote internal development and to meet the needs of a nation's people. This may preclude the development of an insurgency. However, if an insurgency does develop, IDAD is also a strategy for counterinsurgency activities.

Participation by US government agencies in any action taken by another government to free and protect its society from subversion, lawlessness, and insurgency is known as foreign internal defense (FID). Military support to FID is provided by the Commander in Chief (CINC) of the unified command in the region. US armed forces can provide resources such as materiel, advisors, and trainers to support the host nation (HN). Joint and combined exercises can serve to show US support for the HN's government. In extreme situations, US combat forces may be directly employed to support another nation's counterinsurgency operations. US combat operations under these circumstances establish conditions that permit the HN to regain full control of counterinsurgency operations. Activities under the auspices of FID may serve one or more purposes:

- Development of sustainable capabilities or institutions within the HN.
- Improvement of the people's quality of life within the HN.
- Increase of the HN's capability to provide for its own security.
- Improvement of the standing of the HN's military (as well as the US military and the US government as a whole) in the eyes of its people.

Together these factors contribute to the legitimacy of the HN's government and promote stability. Stability is advanced in the country and the entire region. FM 100–20 provides a detailed description of US foreign-assistance programs. Nation assistance activities also contribute to local and regional stability by assisting the development of sustainable institutions that meet the needs of the HN's people. Chapter 2 addresses nation assistance in detail.

When supporting counterinsurgency operations, force protection is a critical activity. It requires a detailed threat-level analysis of the area of operations. This analysis must be updated throughout the operation. During an insurgency, the threat level may vary from one location to another within a country or region. The insurgents maybe drawing their support from a particular area, ethnic group, or social class (farmers, miners, and so forth). The threat may also vary from one time to another. The relative power of the insurgents may wax and wane based on popular opinion. Changes in the political or economic situation or even the season of the year may also affect their strength. A continuous threat–level reassessment is essential for maintaining the correct defensive posture.

Intelligence relating to the local threat may be obtained from a unit's higher headquarters, other units operating in the same area, or the local police. HN forces may provide the most current intelligence about the area of operations. Planning by US forces is conducted based on current and projected capabilities of insurgent organizations. Even in an apparently benign environment, planners establish contingency plans. Engineer soldiers may become targets for insurgents due to the dispersed mode in which they normally perform their missions. They are particularly vulnerable during deployments for construction projects at remote locations. Engineer equipment and logistics parks are large and difficult to secure or defend. They present easy targets for insurgents. Soldiers operating equipment or hauling materials are vulnerable to ambush by fire, land mines, or booby traps. To protect the force, leaders must establish sound operating procedures before deployment. They must reinforce them throughout the operation. Soldiers should be trained in common-sense countermeasures. If appropriate (and the political situation permits), soldiers may wear helmets and load–bearing equipment and carry weapons to present a "combat–ready" appearance. If a unit appears to be professional, competent, and prepared, it will create the perception among the local populace that the unit is not an easy target. Measures should include such things as—

- Knowing the nature and degree of the local threat.
- Maintaining a low profile when in public (off-duty time).
- Traveling in groups when off duty.
- Observing standard operations security (OPSEC) procedures.
- Reporting suspicious packages, vehicles, or individuals.
- Restricting the release of personal data (to include itineraries) on key personnel.
- Providing adequate job-site security.

An insurgent threat is similar to a terrorist threat, particularly if an insurgency is in an advanced stage. Detailed guidelines for force protection under these conditions are provided in Chapter 5 and in Appendix A.

An issue related to force protection is the promulgation of rules of engagement (ROE). ROE are established and promulgated by the headquarters responsible for an operation. In some cases, the National Command Authorities (NCA) will establish the ROE. The unified command responsible for the area in which the operation is being conducted will then disseminate them. ROE must be clearly stated by the chain of command. All soldiers participating in the operation must understand them. ROE should be committed to memory; each soldier must be prepared to act properly in situations without having to first review his ROE instruction. This is essential to prevent incidents resulting from improper use of force. Such incidents could result in unnecessary injury or death of a local national due to lax enforcement of the ROE. At the other extreme, misunderstanding could cause the injury or death of a US soldier due to failure to take appropriate and prudent action.

Engineers participating in counterinsurgency operations may support the internal development of the HN, its internal defense, or both. They can provide mobile training teams (MTTs) or advisors to the HN, take part in single-service deployments for training, or participate in joint or combined training exercises. In an active insurgency, countermine operations maybe a major factor. The degree of importance of these operations will depend on the insurgent's tactics and the HN's current capabilities.

While supporting a counterinsurgency, engineers may operate as part of a joint or combined force or as an independent unit. They may be augmented with combat support and combat service support assets. Engineers supporting an Army combined-arms unit will operate under the command and support relationships specified in the pertinent operation order (OPORD). They will receive missions and support as outlined in FM 5–100. Command and support relationships for Army engineer units participating in multiservice engineer exercises will be specified in the OPORD for those exercises.

Army engineer units operating independently may be placed in any of several command or support relationships and receive support from several sources. Actual relationships in effect during an operation will be specified in the OPORD for that operation. Engineer teams, detachments, platoons, or larger units may report directly to the US security assistance organization (SAO) in a country. Alternatively, these units may be directly attached to, or under the operational control of, a joint task force (JTF) (such as JTF–Bravo in Honduras) already in the HN. For larger engineer exercises, an engineer battalion, group, or brigade headquarters may provide command and control for engineer units participating in the exercise.

When an engineer unit deploys independently, it will usually be augmented by combat support and combat service support assets. These units meet logistical, medical, personnel, communications, and other requirements that cannot be met internally. If the situation in the area of operations warrants, combat assets may also be attached for security augmentation. In this case, infantry or military police may provide security where engineer soldiers would otherwise have to provide their own. The engineer unit may receive attachments of other engineer assets to provide a capability not available in the unit. This support element will be tailored to the needs of the entire engineer task force. Considerations include the-

- Size of the engineer task force (including support assets).
- Assets organic to the engineer unit.
- Support available from the HN and US assets already in the HN.
- US Army Corps of Engineers (USACE) area or resident engineer officers.
- Distance from in-country support to the deployment site.
- Security provided by an HN, freeing US engineer soldiers to concentrate on construction projects.
- Assets required to develop the deployment site.
- Mission-specific requirements such as drilling a well or providing medical support for the local populace.
- Climatic conditions that may affect water, maintenance, equipment, or other requirements.
- Security requirements based on the threat's activity level and capabilities.

When an engineer headquarters deploys, it will normally control the assets that augment the task force.

An engineer unit's staff must conduct detailed planning to support augmentation for a successful operation. If a deployment is to be conducted by a platoon or company, the next higher headquarters must be involved in the planning process. Planners should use their past experiences in similar situations and that of other units, through after-action reports, lessons-learned reports, and articles published in professional magazines.

Engineers operating independently (individuals, teams, companies, battalions, and so forth) must become familiar with several agencies and organizations. While operating in the HN, they must conduct close and continuous coordination with them. They include the–

- Country team at the American embassy (which also acts as a tie-in to the HN's national government).
- Local government in their area of operations.
- HN military forces in their area of operations.
- HN police forces in their area of operations.
- US SOF or conventional forces located within the HN.

Chapter 2 contains a detailed description of the country team concept. It also explains the SAO, which is part of a country team.

When required, a foreign internal defense augmentation force (FIDAF) may augment the SAO. The FIDAF is a composite organization operating under a US unified command or a JTF. Their missions include assisting SAOs with training and operational advice. They also provide assistance to HN forces. There is no set structure for the FIDAF-it will be tailored to the HN's needs. Engineer participation in a FIDAF may range from being part of a FIDAF staff to providing augmentation with engineer MTTs or units (detachments or larger organizations). FM 100–20 provides a detailed explanation of the FIDAF concept.

Army engineers involved in civic action projects may interact with the United States Agency for International Development (USAID), which is part of the country team. USAID is primarily concerned with developmental assistance as well as humanitarian and civic assistance (HCA). It supervises and gives general direction on all nonmilitary assistance programs under the Foreign Assistance Act of 1961, Public Law (PL) 480 (Food for Peace), and similar legislation. It administers HCA programs in conjunction with the US Department of Agriculture. Engineers may support HCA projects during either joint exercises or single-service deployments for training. USAID coordinates these projects to ensure that they support the overall US assistance plan in the HN.

Typically, military engineers work in close cooperation with civil affairs organizations and programs. The programs' objectives are to mobilize and motivate citizens to assist the government and military forces. Constructing communication links is a principal means of spreading government influence and encouraging national unity in developing countries. A system of roads, airstrips, and wharf facilities allows services such as health, education, and welfare to reach the population and encourages central marketing of rural produce. Such development should be a governmental priority with suitable projects identified as civil affairs operations.

Engineer units must coordinate closely with the local government in their area of operations. This coordination improves civil–military relations and reduces the chance of misunderstanding between US forces and the local populace. It also assists US forces in locating resources and reinforces the objectives of FID activities. The local government is best suited to address the particular needs and desires of the populace. US engineers must remember that they are guests in the HN. They are there at the invitation of that country's government to provide cooperative assistance.

Joint action by US engineers and the HN's military and civilians reinforces the concepts of US–HN cooperation and the transferal of capabilities from the US to the HN. Civic action projects, in particular, should be joint ventures. When US forces and HN personnel work together on a project, several benefits are gained:

- The local populace gains ownership of the finished product.
- The HN military and civilians gain a capability (management, construction skills, and so forth).
- The fact that the HN government is helping its society is reinforced in the eyes of the people.
- Perceptions regarding the US are improved.
- US soldiers gain a better understanding of the HN.

Coordination should be conducted with local HN military and police forces regarding security, intelligence, and combined operations. Security for US engineer operations should be provided by HN forces whenever possible. This applies to security on job sites, in base camps, and during convoys. The HN presence serves several purposes:

- It reinforces the concept of cooperation between the US and the HN.

- It reinforces the perception in the eyes of the local populace that the HN is the lead agency in the operation.
- It prevents misunderstandings due to differences in language or customs.
- It reduces negative publicity that may result from direct confrontation between US forces and the local populace.

If an engineer unit is operating independently in the same country as SOF, or other conventional units, coordination (and if possible, liaison) should be established. In some cases, the engineer unit will receive support through a unit that is already in country (for example, through JTF–Bravo in Honduras). SOF located in the HN, specifically Army SF, civil affairs (CA), and psychological operations (PSYOP) units, may provide the engineer unit with essential information. This includes the customs, traditions, and needs of the local populace. SOF may also be able to provide current intelligence, backup linguistic support, and special skills to support engineer operations. These SOF units are regionally oriented and have a great deal of expertise in their areas of responsibility.

The types of support that engineers may provide include a variety of activities. Those listed below were described in Chapter 2. The only difference in application is the environment in which they are conducted and the commensurate increase in the threat level.

- MTTs.
- Technical assistance teams.
- Engineer advisors.
- Joint exercises.
- Engineer-unit deployments for training.
- Military civic action.

In addition, engineer support for counterinsurgency operations may include support for SOF or combat operations.

Some SOF units, such as SF units, have organic engineer personnel. There are two military occupational specialty (MOS) 18C engineers per 12–man operational detachment A. These 18Cs are well–trained in demolitions but have limited construction expertise. FM 31–20 and branch-specific manuals provide doctrine relating to these personnel. During counterinsurgency operations, conventional engineer forces may be tasked to support SOF. Engineers may provide combat, sustainment, and topographic engineering support.

SOF survivability can be enhanced by protective structures that harden critical facilities at SOF operational bases. Engineer units can provide technical assistance in the use of natural and artificial camouflage measures to conceal SOF activities. They can also install protective barriers that support base defense plans. In addition to these survivability tasks, engineer units can also construct full-scale target mock-ups and rehearsal sites and perform other tasks that support SOF sustainment and premission training.

Topographic engineer support may be provided to SOF in the form of standard or special products. Maps, terrain–analysis products, and digital terrain data are provided to the SOF commander so he can develop plans that make the best use of terrain. If available, engineer terrain detachments may provide assistance to SOF in the intelligence preparation of the battlefield (IPB) process.

During the third and final stage of an insurgency, also known as a war of movement, HN forces are engaged in a conflict with organized insurgent forces. Operations at this time closely resemble a conventional war, although the insurgents may continue to employ guerrilla and terrorist tactics on a localized basis. US engineer operations during a war of movement will focus on support to US and HN military forces. Combat engineer units will provide mobility, countermobility, and survivability support, while other engineer units provide sustainment and topographic engineer support. FM 5–100 and other related doctrinal manuals apply during a war of movement.

CONSTRUCTION DURING COUNTERINSURGENCY OPERATIONS

Construction support for counterinsurgency operations may be performed as exercise-related construction (ERC) training projects or civic action projects. ERC supports Chairman, Joint Chiefs of Staff (CJCS) exercises outside CONUS. Joint Staff ERC funds are expended on enduring improvements and structures constructed to directly support these exercises. Water wells, airfields, and so forth, may be constructed to support deploying forces. ERC funds are used for project materials and project-specific petroleum, oils, and lubricants (POL). Transportation costs associated with movement of personnel and equipment from their home installation to exercise sites and back is also provided by the Joint Staff. ERG projects are normally accomplished through troop construction. Projects may be accomplished by contract construction, however. Other costs associated with exercises come from operation and maintenance (O&M) funds except as specified below.

Some CJCS exercises may be conducted specifically to train US engineer forces. These engineer training exercises are undertaken to-

- Provide United States Army Reserve (USAR) and Army National Guard (ARNG) engineer units the opportunity to deploy to a remote location to fulfill annual training requirements.
- Enhance the readiness of participating US and HN units.
- Expose US active and reserve component units to a bare-base and unusual training environment.
- Develop a positive image in the HN toward the US government and its armed forces.
- Foster better military-to-military relationships.
- Evaluate logistical sustainment operations.

Funding for materials and project–specific POL for these training projects is generally provided by the HN in recognition of its receipt of a finished road, bridge, or other product. The US expends funds on these projects because they are training opportunities.

Another source of funding for construction is Section 401, Title 10 US Code, which provides funds for HCA. These programs are authorized by statute to assist the HN through—

- Medical and dental care and preventive medicine and veterinary services provided in a country's rural areas.
- Construction of rudimentary surface transportation systems.
- Well drilling and construction of basic sanitation facilities.
- Rudimentary construction and repair of public facilities.

Congress authorizes the use of O&M funds for HCA. These funds cover materials, fuel, and equipment leases directly related to an HCA project. HCA may not be provided to any individual, group, or organization engaged in military or paramilitary activity. Projects proposed for HCA are reviewed and approved by both the DOD and the Department of State before initiation. Although these projects are normally planned for construction during CJCS exercises, they may also be constructed during single-service deployments for training.

Projects constructed as either training missions or as civic action projects should—

- Meet national developmental criteria.
- Cause no offense to cultural or religious norms.
- Provide a low–technology end product.
- Require minimal maintenance.
- Make maximum use of local resources.
- Require a minimum US troop deployment.
- Be achievable in a short time frame.
- Provide maximum visibility and credit for local government.
- Be constructed in favorable seasonal conditions.
- Have a measurable impact.

These projects may make some use of local contractor support, for example, for the provision of construction materials. This has two effects. First, the material or service the contractor provides contributes to the finished construction project that is an asset to the local populace. Second, the ability of the HN to conduct its own development is increased as contractors gain experience. An engineer leader involved in contracting activities requires expertise or training as a contracting officer's representative or, as a minimum, should have ready access to contracting officers and legal advice.

PEACETIME TO COUNTERINSURGENCY TRANSITION

It is possible that a nation to which the US has provided peacetime assistance may become the target of an insurgency. If this occurs, the NCA, with the advice of the CINC of the appropriate unified command, may shift the focus of US–forces assistance activities. Specific economic or social conditions that the insurgents are exploiting maybe targeted for correction. If a particular area of the country becomes a base of operations for the insurgents because it is inaccessible, the construction or improvement of surface transportation systems in that region may be appropriate. The goal of these activities must be to isolate the insurgents from the populace, while strengthening the bond between the people and the government of the HN. Planning for this transition from a peacetime to a counterinsurgent focus must be a joint effort with complete participation by the HN, the US country team, and the CINC of the unified command.

Chapter 5

Combatting Terrorism

The Marine deployment in Beirut was not designed to be an antiterrorist campaign. It never became such a campaign, although the Marines themselves became targets of terrorists. The Marines did however expend substantial effort in defending their own positions against attacks from conventional and later from terrorist forces.

Technical Memorandum 12-89, Military Forces in Urban Antiterrorism
US Army Human Engineering Laboratory
Aberdeen Proving Ground, Maryland

Royal Ulster Constabulary stations were protected by heavy fortifications, including steel and concrete blockhouses, car barriers, high mesh fences to detonate rockets, controlled entryways, blast resistant walls, steel gates, and armed gatemen. In troubled areas, the stations were fortified with stone walls, antimortar fences, tall gates, floodlights, steel window shutters, closed circuit television surveillance, and careful inspection of all incoming vehicles.

Technical Memorandum 12-89, Military Forces in Urban Antiterrorism
US Army Human Engineering Laboratory
Aberdeen Proving Ground, Maryland

Terrorism occurs across the operational continuum. If the terrorist threat to United States (US) interests increases anywhere in the world, the US military can be a priority terrorist target. This is because the US military has great visibility and symbolic value. When operating in a low intensity conflict (LIC) environment, leaders at all levels must be aware of the potential terrorist threat.

POLICIES AND RESPONSIBILITIES

The Department of the Army (DA) policy for combatting terrorism is to-

- Prevent terrorism through protective and preventive antiterrorism measures.
- Respond forcefully with properly trained personnel when preventive measures fail to quickly terminate incidents on Army installations.

DA is responsible for protecting its own personnel. It also provides technical assistance, or forces, when directed. However, the Department of Defense (DOD) is not a lead agency for combatting terrorism. The Department of State is responsible for dealing with terrorism that involves Americans abroad. It is also responsible for handling foreign relations aspects of domestic terrorist incidents. The Department of Justice (DOJ) deals with domestic terrorism. Investigative and operational responsibility rests with the Federal Bureau of Investigation (FBI).

DOD personnel become involved with the lead agencies when planning for and dealing with terrorist incidents. Outside the US, contingency plans are developed and activities conducted according to Status of Forces Agreements (SOFAs). These are coordinated with the Department of State. Within the US, the Posse Comitatus Act (18 US Code 1385) restricts the use of military personnel. Except as authorized by statute or the Constitution, the act prohibits DOD personnel from assisting local law enforcement officers as part of a *posse comitatus*. It also prohibits using troops to execute federal laws. Policies and procedures for Army assistance to the FBI within the US are outlined in the following:

- Army Regulation (AR) 500-1.
- AR 500–50.
- DA Civil Disturbance Plan (Garden Plot).
- Joint DOD/DOJ/FBI Memorandum of Understanding (MOU) entitled "Use of Federal Military Force in Domestic Terrorist Incidents."

Within Headquarters, DA, a joint team has been formed by the Deputy Chief of Staff for Operations and Plans (DCSOPS) and the Deputy Chief of Staff for Intelligence (DCSI). It is called the Antiterrorism Operations and Intelligence Cell (ATOIC). The ATOIC is located in the Army Operations Center (AOC). The cell monitors terrorist activities worldwide 24 hours a day. It gathers terrorist threat information from all sources and interprets, analyzes, and disseminates it. The ATOIC-

- Coordinates with other government agencies.
- Provides a daily memorandum to senior Army leaders.
- Sends appropriate messages to commands.
- Provides resources and necessary support during crisis response operations.

Engineer units combat terrorism in three ways:

- Terrorism countermeasures in engineer units (force protection).
- Support for antiterrorism (defensive) operations.
- Support for counterterrorism (offensive) operations.

TERRORISM COUNTERMEASURES WITHIN ENGINEER UNITS

Engineer units operating in a LIC environment are normally under a unified command. That command includes threat assessment in its operational planning. It incorporates the terrorist threat in the "enemy" subparagraph of its operation orders (OPORDs). OPORDs specify any unique force protection measures required by the threat level. The command continuously evaluates the terrorist threat throughout an operation. It updates subordinate units on the current threat condition (THREATCON). AR 525–13 includes a detailed description of THREATCONs and response measures.

Engineers may become targets for terrorists due to the mode in which they perform their missions. They are particularly vulnerable to terrorism when deployed on construction projects. Such projects generally require that they operate over large areas with outlying work sites. Engineer equipment and logistics parks are large and difficult to secure or defend. They present easy targets for terrorists. Soldiers operating equipment or hauling materials are vulnerable to ambush by fire, land mines, or booby traps. To protect the force, leaders must establish sound operating procedures before deployment and reinforce them throughout operations. On the job, soldiers should—

- Check for booby traps during preventive maintenance checks and services (PMCS).
- Vary logistical supply routes and schedules.
- Vary travel routes to and from work sites.
- Vary daily schedules.
- Prevent direct access by the public to sensitive areas.
- Equip sensitive areas, such as arms rooms, with duress alarms.
- Restrict the release of personal data (to include itineraries) on key personnel.
- Report suspicious incidents, individuals, vehicles, packages, and so forth.

Work-site security requires a force capable of discouraging a terrorist attack. During construction operations, it may not be practical to arm all soldiers at all times. It may even be undesirable when working with local civilian personnel. There are several options that may be used to secure the work site. The option selected will depend on the THREATCON and the local situation. These are a few of the options:

- A designated element within the engineer force can maintain security. This security element is armed and carries ammunition.
- A separate US element can provide security for the engineer force.
- A host-nation (HN) element can provide security for the force.

Physical security for construction missions can be enhanced by security patrols and consolidating construction equipment in planned, safeguarded areas at the end of the workday. Use sentries and observation posts under the supervision of a duty officer for night security. Establish a reaction force under the duty officer if the THREATCON warrants.

The security of troop living areas, headquarters elements, weapons storage areas, communication assets, and so forth, is a key planning consideration. Security planners must prioritize these assets when developing the unit's force-protection plan. They must balance the benefits of asset dispersion and the cost of an increased guard requirement.

Leaders must develop personnel security by educating the chain of command down to the lowest level. Simple defensive measures, coupled with a general threat awareness and an understanding of terrorist techniques, reduce soldier vulnerability. Terrorist tactics include-

- Bombing.
- Arson.
- Hijacking.
- Assassination.
- Ambush.
- Kidnapping.
- Hostage taking.
- Robbery.
- Maiming.

- Raids (armed attacks).
- Seizure.
- Sabotage.
- Hoaxes.
- Use of nuclear, biological, and chemical (NBC) weapons.

Bombing is the tactic most commonly used by terrorist groups. Bombs are popular weapons because they are cheap to produce and easy to make. They are adaptable to a variety of uses and difficult to detect and trace. Terrorists use a variety of methods to deliver bombs, including vehicle bombs, laid charges, projected bombs, and postal bombs. They may be activated by command detonation, action of the subject or target, or by timer delay.

Soldiers should be trained in common-sense terrorism countermeasures, such as-

- Knowing the nature and degree of the local threat.
- Maintaining a low profile when in public (off-duty time).
- Traveling in groups when off duty.
- Observing standard operations security (OPSEC) procedures.
- Allowing access on an "as required" basis to camps, clubs, facilities, and so forth.
- Reporting suspicious packages, vehicles, or individuals.

The best protection against terrorism is a well-disciplined unit made up of fit, well–trained, and alert soldiers. All soldiers should consider terrorism routinely in the conduct of their day–to-day activities.

ANTITERRORISM OPERATIONS

When deterring terrorism, engineer units may support US military forces, nonmilitary governmental agencies, or government agencies of other nations. The units use active and passive defensive measures known as antiterrorism operations. Antiterrorism operations include all measures that installations, units, and individuals take to reduce the probability of their falling victim to a terrorist act. These operations include collecting and disseminating timely threat information, conducting terrorism–awareness programs, and implementing defensive measures. Because absolute protection against terrorist activities is not possible, protective plans and procedures reflect the specific threat in the area of operations. Protective measures should strike a reasonable balance between the protection desired, mission requirements, and availability of manpower and financial resources.

Likely engineer missions in support of antiterrorism operations include mobile training teams (MTTs) for HNs and training for US personnel. Engineer personnel may be sent outside the US as MTTs to conduct force-protection instruction for other governments. These teams consist of officers, senior noncommissioned officers, US Army Corps of Engineers (USACE) civilian engineers, or a combination thereof. Foreign governments request MTTs under the auspices of the International Military Education and Training Program (IMETP) or through foreign military sales (FMS). Requestors obtain MTTs through the in-country US security assistance organization (SAO). The teams provide the requesting government a self–training capability by training selected HN personnel. These trainees then constitute an instructional base.

Training from engineer MTTs ranges from classes on wire obstacles and bunkers for base camps to design and construction of protective barriers for installations. Chapter 2 contains additional information on MTTs.

Engineer personnel may provide assistance in the design or construction of protective structures for US military facilities worldwide. Engineer troop units may construct permanent or temporary antiterrorist measures for US forces. They may also provide training on these activities.

USACE is responsible for developing, establishing, and maintaining protective design criteria and standards for Army facilities. Army military construction (MILCON) policy states that protective design criteria will be considered for all proposed Military Construction, Army (MCA) projects. These criteria apply to all major construction in the continental United States (CONUS) and outside continental United States (OCONUS). The long–term threat will drive facility planning and design enhancements. When combatting terrorism, facility planning and design are based on threat targeting patterns, weapons, and delivery systems deemed probable for the next ten years. Department of the Army Military Construction Programming Policy provides specific procedures for establishing and validating the threat.

For existing facilities, security construction projects increase their antiterrorism posture. Measures are employed to reduce specific vulnerabilities to identified threats. It is important to understand the purpose of these measures. They protect assets within the facilities, not the facilities themselves. The facilities are part of the protection. Some possibilities include–

- "Hardened" concrete guard houses with bullet–resistant glazing.
- Hydraulically or manually operated vehicle crash barriers.
- Personnel gates designed to limit the number of personnel passing through at one time.
- New or increased perimeter and interior security lighting.
- New, improved, or repaired security fencing.
- Intrusion detection systems.

The Provost Marshal's Office (PMO) Physical Security Section and the Directorate of Engineering and Housing (DEH) should plan these projects jointly. The PMO provides an assessment of terrorist capabilities while the DEH provides engineering expertise. USACE may provide designs, or the PMO, DEH, and local contractors can develop them. Depending on availability of funds and statutory requirements, contractors or troop units can construct the projects.

Engineer units may construct expedient barriers to combat terrorism. These barriers may protect temporary facilities, such as base camps, or permanent facilities where the terrorist threat has increased in activity or technology level. USACE is responsible for standard, expedient structure designs. Use ingenuity in the absence of standard designs. For example, parking a bulldozer in front of an unused vehicle gate at night reduces the chance of a terrorist's vehicle crashing the gate.

COUNTERTERRORISM OPERATIONS

Counterterrorism operations are offensive actions taken in response to a specific terrorist threat. Counterterrorism response measures are normally conducted by specially organized forces trained and equipped for the mission. These include preemptive, retaliatory, and rescue operations. Support by conventional engineer forces is limited.

Engineer units may support counterterrorist operations by building rehearsal sites. This requires a detailed intelligence analysis and an understanding of how the site will be attacked. Intelligence often provides only a limited knowledge of the location through photos, human intelligence (HUMINT), and so forth. Engineer planners must interpret this intelligence. They then design an appropriate rehearsal site based on sound engineering judgement. The rehearsal site must accommodate force-projection means (helicopters and so forth) and live fire exercises.

Topographic engineer units provide support, such as city maps and special map products to military and civilian counterterrorist units. Data–base availability for the area of interest and the proximity of a topographic unit impact how quickly units can support a specific operation.

An engineer unit may act as part of a reaction force or may help secure an area where a terrorist incident is occurring. These situations may occur as part of a general scenario when the engineer unit is the largest US force in an area. Examples would include an engineer base camp during a Central American deployment or an engineer unit whose commander is a base commander under the area-support concept.

Chapter 6

Peacekeeping Operations

The French engineer force, as part of the French element of the Multinational Peacekeeping Force, was initially tasked to clear mines, debris, and fortifications from a large part of the city center. The density of mines and booby traps was exceptionally higher than that found in the mining doctrine of any European army. The level of danger was extremely high and the need to be physically and mentally prepared to deal with the stress became apparent.

French Army Combat Engineer Experience in Beirut, Lebanon

Engineers who are part of a peacekeeping force must prepare to operate in a tense environment. These operations are inherently stressful due to the peacekeeping force's position between two (or more) armed belligerents. During these operations, engineer missions cover a broad spectrum from facilities construction to minefield clearing.

PLANNING CONSIDERATIONS

Specific planning guidance is provided for each peacekeeping operation. Memorandums of instruction (MOIs) are prepared by the major organization providing units and elements for a US peacekeeping contingent. These MOIs contain information from the terms of reference that govern United States (US) military participation in the peacekeeping operation. MOIs provide units with information needed for preparation, deployment, and execution of their mission. Each MOI should contain information on these topics:

- Organization and equipment.
- Operations.
- Intelligence.
- Personnel.
- Logistics.
- Public affairs.
- Finance.
- Air operations.
- Nuclear, biological, and chemical (NBC) defense.
- Command relationships.
- Communications and electronics.
- Rules of engagement (ROE).

The threat level in the area of a peacekeeping operation is a major consideration when planning and executing the mission. The engineer commander (or senior engineer leader) must keep abreast of the current threat level in his unit's area of operations. The threat may come from regular forces on either side of a dispute or from other parties who want the peace effort to fail. The threat depends on many factors, including the-

- Number of factions affecting the situation.
- Various ideologies involved.
- Sophistication of the belligerents' weapons.
- Specific techniques of the disputing parties.

The threat level also depends on how impartial and effective the belligerents perceive the peacekeeping force to be. Threat information should be relayed up and down the chain of command. Often, the peacekeeper on the ground is in the best position to observe signs of change in the threat level. This information is passed up the chain of command for integration into the overall threat assessment.

Physical security measures protect the force and ensure that it can complete the mission. Security requirements for force protection vary depending on the threat level in the area of operations. Planning is based on current and projected capabilities of the disputing parties. Even in a seemingly benign environment, planners prepare force contingency plans and emergency plans. Countermeasures are developed for use against likely threat operations. Training is conducted to ensure an appropriate and swift response. Overall security measures are critical to a force constrained to set up and operate within a limited area for an indefinite period of time.

Personnel security is paramount in peacekeeping operations. Leaders must educate soldiers in threat tactics and recognition through a continuing awareness program. ROE are developed for each peacekeeping operation. They should be clearly and precisely stated in the mandate establishing the peacekeeping force. The ROE describe when peacekeepers may use force to resist attempts to prevent them from performing their duties. The right to use force in self-defense is never restricted. The chain of command must ensure that all echelons involved in the operation understand the ROE. Units that will take part in peacekeeping operations will establish training programs. As a minimum, training should include-

- Peacekeeping operations.
- ROE.
- Legal considerations.
- Medical threats.
- Field sanitation and hygiene.
- First aid.
- Map reading.
- Weapons and equipment identification.
- Mine identification and clearance procedures. (You may contact the US Army Engineer School Hotline for assistance (AUTOVON: 676–7324).)
- Operation of non–US construction equipment that may be present in the area of operations.
- Culture, language, habits, religion, and characteristics of the local people.
- Survival in the environment of the particular area.
- Special skills (for example, NBC defense, desert operations, and jungle operations).

If engineer soldiers will be integrated with other branches conducting peacekeeping operations, they should also be trained in patrolling, operating checkpoints and observation posts, and air assault operations. Engineer units should train on military operations on urbanized terrain (MOUT) techniques if urbanization is present in their area of operation.

The size and type of engineer unit that a peacekeeping force requires to support a peacekeeping operation depend on several factors. These include the-

- Specific type of peacekeeping operation.
- Support responsibilities outlined in the terms of reference.
- Size of the force being supported.
- Duration of the operation (or rotation within an operation).
- Environmental considerations, such as the degree of urbanization or the presence of mine warfare.
- Degree to which the belligerents are maintaining peace.
- Existing facilities and services.
- Availability of contracted engineering support.

The terms of reference outline the specific missions of the US contingent to a peacekeeping force. In some multinational operations, another country may be tasked to provide engineer support to the force as a whole. In other cases, the US may be tasked to provide all engineer support for the operation. This type of support may include base development and maintenance. There may also be combat engineering missions that affect the whole force. Individual country contingents (to include the US) still require some internal engineer support to implement force-protection measures and to conduct possible combat engineering missions.

The size of a US force provided to a peacekeeping operation may range from several observers, through an infantry battalion reinforced with support assets, to an even larger force. The size and composition of the supporting engineer unit varies depending on the specific tasks it must perform. If the force moves into an area with no facilities, the requirement for construction engineering skills depends on several factors. These include whether facilities will be constructed by the US force, by another country, or by contract. If the peacekeeping force moves into existing facilities, the requirement for construction engineering skills depends on who will maintain the facilities. This could be US forces, another contingent's engineers, or a contractor.

How the belligerents comply with cease-fire agreements affects the need for combat engineering skills. A relatively benign environment requires minimal combat engineering support. This is the case with the Multinational Force and Observers in the Sinai, where each US infantry battalion is supported by a reinforced engineer squad during its rotation. In more threatening environments (where all disputing parties are not complying with a cease-fire, for example), the need for combat engineering skills increases. In some cases, one or more belligerents may continue to conduct mining operations, for example, in contested areas or along peacekeeping–force patrol routes. They may emplace booby traps as defensive measures or to harass opponents or peacekeepers. The belligerents may conduct openly aggressive activities such as ambushes or raids. In any of these conditions, engineer soldiers may be required to conduct standard combat operations, possibly under fire. Due to the high–risk and high-stress nature of these operations, the number of combat engineers supporting the force should be increased.

Specific engineer missions will fall into one of two categories: sustainment engineering or combat engineering support.

SUSTAINMENT ENGINEERING

Sustainment engineering includes tasks that support the force through the construction and repair of billeting, support, and logistical facilities as well as lines of communication (LOC). These tasks may include the construction, maintenance, and operation of water, electrical, and sanitation utilities. Sustainment engineering also includes locating water sources and, if necessary, drilling wells for water supply. Engineer unit capabilities vary, depending on the specific type of unit. Planners must determine their specific engineering requirements. They can then request the appropriate types of units based on Field Manual (FM) 5–100 and Appendix C of this manual.

Sustainment engineering provides an adequate support base for the peacekeeping force. This base must provide living conditions that are secure, healthy, and as comfortable as possible for the peacekeeping force. It must also provide enough administrative and maintenance space for units supporting the force and for secure storage of all supplies and material. The support base's size depends on the size of the peacekeeping force and its supporting elements. Specific requirements may also affect the base's size. For example, if rations are delivered twice a week versus daily, ration storage facilities are designed accordingly. Water storage requirements are affected by the availability of potable water, the size of the force, and the local climate.

Command and control of engineer units involved in sustainment operations vary from one situation to another. Command relationships and control procedures must be established before beginning the operation. In cases where a small engineer contingent (such as a squad) is attached to a unit, it should be further attached to a subordinate headquarters. This subordinate unit will support the engineers and allocate their efforts to support the whole force. For example, an engineer squad supporting an infantry battalion may be attached to the headquarters company if the company commander is also the base-camp commandant. If a large engineer organization (company or larger) supports a force, the engineer cormmander should recommend command or support relationships to the force commander.

The peacekeeping-force commander and the engineer planner must establish parameters for base development. The threat level determines the defensive measures which will be taken to protect the peacekeeping unit. This affects the types of structures and construction techniques used by engineers in base development. An area designated as a base of operations may be protected by a range of means. A system of wire obstacles, antivehicular obstacles, and guards may be adequate in a low–threat environment. Bunkers, barricades, and continuous patrols may be required in high–threat areas. The expected length of peacekeeping operations impacts the construction techniques and standards used by the engineers. "Tent cities" may be used for a short–term operation. They will be set up with protective structures for personnel and essential facilities. A long–term peacekeeping operation may require more permanent facilities. In this situation, engineers attached to the peacekeeping force may continuously upgrade facilities if mission requirements and fiscal resources permit.

Logistical support may come from several sources, depending on the specific peacekeeping operation. Logistical support information is provided in the letter of instruction (LOI) published by the headquarters which is planning the peacekeeping operation. Primary logistic support comes from a military logistic-support unit controlled by the peacekeeping command. Civilian contractors may also provide support. Major equipment items may accompany the deploying unit, or the peacekeeping unit may provide them in the area of operations.

Specific engineer sustainment missions include-

- Base-camp construction. Base camps are constructed for billeting the peacekeeping force and as a base of operations. The projected force size, including all support elements and related activities, must be considered when designing the camp. In some cases, personnel from other contingents in the peacekeeping force may be included in the planning figures. Possible changes in the camp's population should be considered. Space for enlarging the camp to meet changing requirements should be provided, if possible. Planners must consider power requirements for the base camp. Tactical generators, a prime power plant and team, the host nation (HN), or a combination of these sources may provide power for the camp. Specific types of structures and physical-security considerations in the design are based on the parameters previously addressed. In some cases, the Army facilities components system (AFCS) may provide designs and bills of material (if the Operations length is less than 24 months). Camp construction may be conducted by engineer troops from the US or other nations or by contractors. US engineers, or those of other countries, may be responsible to oversee execution of a contract for construction.
- Logistics-facilities construction. Logistics facilities are constructed to support and sustain the peacekeeping force. The number and types of facilities vary considerably, depending on the supported force's size and the support functions required. Facility types could include ration storage and issue structures, maintenance facilities, and medical-treatment structures. Bulk–petroleum retention walls; packaged petroleum, oils, and lubricants (POL) storage facilities; and ammunition storage structures may also be needed. Planners must consider power requirements when laying out these facilities. As with base camps, power may come from tactical generators, a prime power plant and team, theHN, or a combination of these sources. Facilities may support only US forces in a unilateral or multinational operation or several nations' contingents in a multinational peacekeeping force. The degree of hardening and other security considerations in each facility's design depends on several factors. These include its function, the sensitivity of its contents, the threat level, and how long it will be needed. As with the base camp, the facilities maybe constructed by engineer troops from the US or other nations or by contractors.
- Lines of communication construction. Depending on where peacekeeping operations occur, US forces may become involved in LOC construction. This situation may occur when peacekeeping forces are positioned between two belligerents in a remote or undeveloped area. The LOC structures, such as roads, bridges, and airfields, may serve several purposes. They may be required for movement of the peacekeeping forces within the area of operations. They may support transportation of equipment and supplies from their sources to base camps and observation posts. Again, the LOC structures maybe built by engineer troops from the US or other nations or by contractors.
- Peacekeeping-specific construction. Several other construction missions are essential for peacekeeping operations. The missions include construction of—

  -Observation posts and towers.
  -Checkpoints.
  -Roadblocks.

These facilities are designed and constructed to meet the same parameters as the base camp.

Observation posts, checkpoints, and so forth, must be clearly marked as peacekeeping–force structures. For example, when these facilities are constructed for a United Nations (UN) peacekeeping operation, the UN flag will be flown 24 hours a day and illuminated at night.

- Potable-water-source development. Water supplied to peacekeeping forces may come from several sources. It maybe provided by Quartermaster water production and distribution personnel or a public water utility or it may be delivered by a contractor. If it is obtained from nonpotable local sources, engineer, quartermaster, and medical personnel have specific responsibilities for water supply. Engineers will locate subsurface water sources (if no surface water sources are available) and construct water wells, if required. (Note: US Army Corps of Engineers (USACE) can provide Water Detection Response Teams who specialize in this area (AUTOVON: 345–2534). These are multiagency teams designed to respond to a request within 24 hours to select well-drilling sites. They use the worldwide Water Resource Data Base, existing publications, maps, and remote sensing data.) Engineers also provide construction support for water–point improvement, if required. Quartermaster personnel select water-treatment and supply-point locations; install purification, storage, and distribution equipment; and treat and dispense water. Medical personnel inspect and approve water sources, define the degree of treatment required, and approve treated water.
- Base and LOC maintenance. This maintenance may be done by a contractor or by military engineer personnel. The terms of reference will outline responsibilities for these functions. Specific missions could include road maintenance and upgrade; culvert replacement; electrical, carpentry, masonry, and plumbing repair; dust control; waste disposal; and other related activities. Prime power teams may maintain power plants and distribution systems. Prime power teams may maintain power plants and distribution systems. When a contractor or engineers from another country are responsible for base maintenance, US engineers may still perform repair and utilities (R&U) activities on a case-by-case basis. This provides flexibility and responsiveness to the commander and training for the engineer soldiers.

COMBAT ENGINEERING

Combat engineering tasks (mobility, countermobility, and survivability) may be conducted by any US engineer unit while supporting peacekeeping operations. Commanders employing engineers should carefully consider the types of engineer units available and their levels of training before committing a specific unit to a mission. Some missions and environments may better suit one type of unit than another. For example, soldiers from an armored division's engineer battalion may be best suited for a minefield-clearing mission (versus a minefield–marking mission) in an urban environment. These engineers may be better suited to conduct the mission than those from a combat heavy battalion or even a combat, corps battalion. However, the lack of a particular type of engineer unit should not prevent the completion of essential engineer tasks.

Command and control of engineer units conducting combat engineer operations depend on the situation, When engineers operate in a low–threat environment, such as training maneuver forces on wire-obstacle construction for force protection, centralized control (for example, general support) is appropriate. When they support a maneuver force in a high–threat environment, such as clearing a minefield or clearing an area of booby traps, a more restrictive command or support relationship, such as direct support or operational control, should be implemented. Chapter 4 of FM 5–100 contains general command and control principles.

Logistical-support channels for combat engineer operations are normally the same as those used in sustainment engineering operations. An exception may occur if an engineer unit is attached to a unit other than that which it normally supports in the peacekeeping force. Logistical support to accomplish engineering tasks will come from the tasking unit if the engineers are attached, operational control (OPCON), or in a direct support relationship to the tasking headquarters. If the engineer unit is in a general support role, logistical support for engineering tasks will come from the engineer unit's parent headquarters. Combat engineer missions specifically related to peacekeeping operations include-

- Construction of command posts and bunkers.
- Construction of force-protection structures such as earth revetments, wire obstacles, and fighting positions.
- Clearance of fields of observation.
- Minefield clearing or marking, to include minefield-fence maintenance.
- Fortification demolition.
- Clearance of mines and debris from roads.
- Clearance of mines and booby traps from buildings, vehicles, and other locations.
- Backup support for explosive-ordnance identification, marking, removal, or demolition.
- Provision of technical expertise to maneuver forces.

Large numbers of antitank and antipersonnel mines may remain in an area when the parties of a dispute disengage. These mines are a hazard to civilians within the area and to the peacekeeping force. US engineer personnel may become directly involved in countermine operations, or they may train others in countermine procedures. Whether minefields are cleared or only marked depends on each situation—the potential threat to civilians and peacekeepers, the risk associated with clearing the particular minefield, and so forth. Engineers performing peacekeeping operations must know the mine-warfare techniques used by the armies that trained the factions in a dispute. The engineers must also learn "local" mine-warfare and booby-trap techniques. Methods of laying and camouflaging mines may vary depending on which side emplaced them. In many peacekeeping operations, a wide variety of mines will be present. Engineers must be trained to identify and clear mines manufactured by the principle arms producing nations.

A peacekeeping situation may deteriorate into an armed conflict between the peacekeeping force and one or more of the disputing parties. If this occurs, the peacekeeping–force commander will try to disengage his unit. If he fails to do this, the peacekeeping force may be involved in combat operations with regular or irregular forces. In these situations, engineer doctrine relating to combat operations is applicable. Engineer units should develop contingency plans and conduct training for this possibility.

TOPOGRAPHIC ENGINEERING

Topographic engineer tasks supporting peacekeeping operations should be tailored to the force, mission, and area of operations. Tasks may include survey, cartography, terrain analysis, and terrain information reproduction.

Specific tasks may include-

- Tactical and civil mobility studies.
- Facility site analysis.
- Communication site surveys.
- Water location.
- LOC planning.
- Weapons emplacement siting.
- Special purpose topographic products to support peacekeeping plans and operations.

Special topographic engineer units can be employed for specific needs. The units may work independently, as part of a US tactical force, or with national, civil, or military forces. FM 5-105 provides the principal doctrine for topographic operations.

Chapter 7

Contingency Operations

Since the [Point Salines] runway was surrounded by construction equipment, bringing in engineer equipment from Fort Bragg was unnecessary. Equipment operators with the first–deployed engineer platoons quickly began using the captured equipment, which included Russian dump trucks and Komatzu bulldozers. As more operators, mechanics and combat engineers arrived, more captured equipment was put into operation.

"Operation Urgent Fury"
LTC Lawrence L. Izzo
The Engineer, Winter 1983-84

Contingency operations (CONOPs) under conditions short of war are politically sensitive military activities normally characterize by the short-term, rapid projection or the employment of forces. This chapter addresses the following types:

- Disaster-relief operations.
- Support to counterdrug operations.
- Security assistance surges.
- Noncombatant evacuation operations (NEOs).
- Rescue and recovery operations.
- Shows of force and demonstrations.
- Operations to restore order.
- Strikes and raids.

The consistent feature of CONOPs is an effort focused on a specific problem (usually in a crisis) and guided at the national level by the crisis action system. This system has six phases and is designed for time-sensitive, joint military planning, Joint Publication (Jnt Pub) 5-02.4 contains details on the crisis action system.

CONOPs under conditions short of war frequently take place far from established military bases of operation. These operations may require temporary development of long lines of communication (LOC) in a hostile environment. They are often undertaken to avoid a crisis or to manage crisis situations requiring the use of military assets to enforce or support diplomatic and informational initiatives. These conditions distinguish CONOPs under conditions short of war from wartime CONOPs, which are often conducted for purely military objectives.

CONOPs may require a wide variety of engineer support. Specific operational requirements may limit participation to a small number of engineers. In other cases, engineer involvement on a large scale maybe necessary. The type of operation determines requirements for combat– or construction–related engineering skills. These skills may be needed simultaneously or sequentially. Large areas of the world where CONOPs may occur are not yet covered by scale maps. The Defense Mapping Agency or Army topographic engineers may provide special quick–response photomaps from satellite imagery, aerial photography, and so forth. Existing host-nation (HN) maps may also be revised or enlarged to meet specific mission requirements. During CONOPs, engineers may work with civilian agencies, such as the Red Cross. Engineers may support combat forces during joint or combined operations. Engineer command and support relationships vary considerably, depending on the circumstances of each specific operation. Logistical support for engineers also varies. Support procedures will be outlined in applicable operation orders (OPORDs).

DISASTER RELIEF

Disaster-relief operations provide emergency assistance to victims of natural or man–made disasters abroad. These operations are responses to requests for immediate help and rehabilitation from foreign governments or international agencies. Disaster relief may include refugee assistance, food programs, medical treatment and care, or other civilian welfare programs. While this section specifically addresses disaster relief, the same principals apply when assisting refugees resulting from other situations such as political turmoil. Army Regulation (AR) 500-60 provides additional information for disaster–relief operations.

In a low intensity conflict (LIC) environment, disasters can worsen already unstable situations. When properly executed, United States (US) participation in disaster-relief and refugee assistance can have significant, positive effects. The military can provide logistical support to move supplies to remote areas. They can locate and extractor evacuate victims as needed and provide emergency communications or conduct direct medical-support operations.

Military elements involved in disaster–relief operations may have a variety of missions. Military personnel assess the damage caused by the disaster and the HN's ability to deal with it. The military executes assistance programs developed by the Department of State or the US Agency for International Development (USAID). Army combat support and combat service support units play a major role in these operations. If needed, combat arms units can provide additional support.

Command and control during disaster-relief operations vary from situation to situation. A command structure may be established for US forces participating in these operations. In some cases, units may work directly for the US country team. Disaster–relief operations are generally joint in nature. Because of the quick response time necessary in disaster–relief operations, it is essential to establish liaison, communication, and operating procedures rapidly. Coordination with HN and international agencies is always essential.

Logistical support from an HN or nearby country maybe provided for disaster–relief operations. A nation that experiences a disaster severe enough to request external support may have difficulty providing for its people. The nation may also be unable to support relief personnel from the US and other nations. Logistical support for disaster–relief operations often flows through LOCs stretching from the US.

Engineers may become involved in disaster–relief operations as individuals, teams, or complete units. Individuals may provide technical assistance to HNs in the areas of damage assessment or engineer work estimation. Teams may provide specialized support such as well drilling, power supply and distribution, or utilities repair or reconstruction. Units may perform these missions:

- Support for search and rescue operations.
- Clearance of rubbled areas.
- Opening of roadways for emergency and medical traffic.
- Restoration of critical facilities, services, and utilities.
- Provision of emergency topographic engineering support.
- Engineer environmental studies.
- Construction of roads.
- Provision of water facilities and distribution.
- Provision of sanitation facilities.
- Construction of displaced-persons camps.

The United States Army Corps of Engineers (USACE) can support disaster–relief efforts by providing expertise through its engineer districts and divisions. Support can include damage survey and assessment teams, contracting support, and technical advice. The US Army Engineering and Housing Support Center (USAEHSC), a field operating agency of USACE, can provide prime power teams and equipment to provide temporary electrical power to key locations and to help restore electrical services. (See Chapter 2 for more information.) Planners must resolve funding issues with USACE early in the disaster–relief response process.

Engineer units supporting disaster–relief operations must take adequate tools and equipment to support all potential missions. For example, a unit deploying to an earthquake disaster area should take all squad and platoon tool kits. The unit should also take specialized tool kits, such as carpenter's, pipe fitter's, and electrician's tool kits. Pioneer-type tools are used to clear rubble, open air holes, assist in entering structures, and so forth. When conducting search operations, specialized tool kits are used to turn off water systems, turn off or seal gas leaks, disconnect electrical systems, and temporarily buttress damaged structures. Construction equipment is used to clear emergency routes, remove rubble, and demolish unsafe structures. When searching partially collapsed structures, cutting torches are needed to cut reinforcing bars. In the same manner, units deploying to a flood disaster area need to consider the mission types they may face based on initial reconnaissance reports. If displaced–persons camps are to be constructed, heavy horizontal and vertical engineer equipment assets will be required. Planners must also design units' support packages (fuel, water, communications assets, maintenance support, and so forth) to facilitate the success of their operation.

When conducting disaster–relief operations, it is critical to maintain personnel accountability. Leaders must keep track of subordinates at all times. Close supervision and personnel training is necessary to prevent the perception of looting. During a crisis situation, a soldier may become separated from his unit and be trapped in a hazardous area. Unit leaders must develop a system to track the location of their soldiers at all times (to include during rest or sleep breaks).

When conducting disaster–relief operations overseas, failure to obtain proper funding sources and authorizations may become a "show stopper" for engineer operations. Immediate efforts must be made to identify find sources and authorizations to procure materials and issue construction contracts.

Another critical factor in disaster-relief and/or refugee operations is to observe ethnic and cultural mores in types of construction and operations. Failure to construct facilities (for example occidental versus oriental style latrines) according to local values may result in unused facilities and/or poor relationships with the local populace.

SUPPORT TO COUNTERDRUG OPERATIONS

Military support to counterdrug operations reduces, as much as possible, the supply of illicit substances available in the US. Military efforts support national drug control strategies and provide an integrated program of counterdrug actions. Military counterdrug efforts always complement, rather than replace, the efforts of other US agencies, the states, and/or cooperating foreign governments. The commitment of military resources is always consistent with US national values and legal framework. The primary legal restraint on military involvement in counterdrug activities is the Posse Comitatus Act. It prohibits military enforcement of civil statutes except in cases and circumstances expressly authorized by the Constitution or Act of Congress.

Military counterdrug support must be a balanced effort to stop the flow of drugs in each phase of the supply cycle: at the source, while in transit, and during distribution. HNs that are source or transit countries receive direct financial and technical assistance. The US government provides operational assistance to HN forces to attack drug-production facilities. The US collaborates with HN law enforcement agencies and armed forces to prevent the export of illegal drugs from those countries. Within the US, military forces can provide support to federal, state, and local agencies (within the restrictions of the Posse Comitatus Act) to locate narcotics sources. An example is locating marijuana plants on public lands.

Military support to efforts to stem the flow of drugs will always be conducted in coordination with federal, state, and local law enforcement agencies. Military support for interdiction includes two operational areas. The first is interdicting drugs in air, sea, and land traffic lanes. The second is detecting, monitoring, and seizing drugs at the nation's borders and ports of entry. Within the US, the military provides support for domestic counterdrug operations to attack the flow of drugs by several methods:

- Military planning and training assistance for domestic law enforcement agencies.
- Equipment loans and transfers.
- Use of military facilities.
- Enhanced roles for the National Guard.

Engineer-specific missions supporting counterdrug operations include-

- Constructing target ranges for law enforcement personnel.
- Constructing temporary operational bases. Facilities may include helipads, fuel storage facilities, maintenance facilities, and billets.
- Producing photomaps of areas where counterdrug operations will take place.
- Constructing or upgrading access roads for drug interdiction patrols.
- Clearing fields of observation for counterdrug teams.

- Locating infiltration tunnels.
- Supporting efforts to encourage farmers to develop alternative crops to those that support the drug trade.
- Rehabilitating existing drug-law enforcement agency (DLEA)-owned buildings to accommodate counterdrug operations or activities, such as an evidence processing facility or operations center.
- Any other mission that is within the capability of military engineer troop units and is also within the legal framework of DOD support to DLEAs.

Army National Guard units have also supported local law enforcement agencies by demolishing buildings used to distribute drugs in the inner city.

During counterdrug operations, engineers must be sensitive to the legal aspects of support to civilian authorities and abide by the Posse Comitatus Act. They must also be aware of the capabilities of the threat. Many narcotics traffickers are well-equipped and heavily armed. The fact that counterdrug operations area peacetime activity must not lull leaders into a false sense of security. Military support to civil authorities in counterdrug operations capitalizes on inherent capabilities of the US military.

SECURITY ASSISTANCE SURGES

The US may accelerate security assistance when a friendly or allied nation faces a threat of imminent harm. Activities are frequently focused on providing logistic support. This includes the provision of additional weapons systems, equipment, and supplies. However, support may include the full range of assistance to include training and financial support. Engineers may aid these surges through mobile training teams (MTTs) designed to transfer skills or to assist in fielding engineer equipment.

NONCOMBATANT EVACUATION OPERATIONS

NEOs relocate threatened civilian noncombatants from locations in a foreign land or an HN. These operations normally involve US citizens whose lives are in danger. NEOs may also include selected HN and third-country personnel, based on the current situation or previously arranged agreements. AR 525–12 outlines responsibilities, policies, and procedures for planning and conducting NEOs.

Military, political, or other emergencies in any country may require evacuation of designated personnel as the situation deteriorates. The Department of State initiates requests for military assistance and obtains necessary clearances from other governments. These clearances can include basing and overflight authorizations and the use of facilities essential to performing the evacuation.

A NEO in a LIC environment usually involves the swift insertion of a force; the temporary occupation of an objective; and a planned, rapid withdrawal. Only the degree of force required for self-defense and the protection of the evacuees is used. Ideally, there should be little or no opposition to an evacuation. In reality, this may not always be the case. A number of factors impact on the specific manner in which forces conduct a NEO. These include the–

- Required speed of the evacuation.
- Potential for violence (threat level).
- Number of individuals to be evacuated.
- Dispersion of individuals to be evacuated.
- Status of ground LOC.

The speed at which an evacuation must be conducted will depend on how rapidly the situation in an HN deteriorates. If intelligence assets provide early warning of the need for an evacuation and the political decision is made to initiate an evacuation, a gradual withdrawal is possible. If destabilization occurs quickly or if a political decision is made to provide an in-country presence as long as possible, a rapid evacuation may be needed. Ideally, dependents and nonessential personnel will have already departed at the direction of the US ambassador before the start of a NEO.

The potential for violence during a NEO depends on the strength of local forces opposing the HN's government. The ability of that government's police, paramilitary, and military forces to maintain order is also key. US forces conducting a NEO may enter a situation where there is little threat from opposing forces and the HN is able to effectively shield US forces from contact with the threat. Sometimes forces opposing the HN may actively attempt to disrupt the evacuation, but the HN is still able to support the NEO. US forces may also conduct a NEO under conditions of strong and heavily armed opposition, with the HN's forces unable to provide any support. In a worst-case scenario, the "hint" nation's forces may actually be part of the threat. In each of these situations, accurate intelligence and detailed planning is essential to ensure a successful operation. The US evacuation–force commander must always be prepared to defend the evacuation effort and provide protection for his forces.

Planning for force protection during NEOs requires detailed analysis of the threat level. Military planners must assess the strength of forces opposing the HN's government and the HN's ability to maintain law and order. Rules of engagement (ROE) must be developed for the operation. They must be understood by all members of the force conducting the operation. Generally, NEOs are conducted during politically sensitive and potentially unstable situations. Local conditions may change rapidly. Commanders should remember that NEOs can quickly turn into peacekeeping operations and plan for these contingencies. All participants in the operation must understand the environment they will enter.

Engineers providing support for a NEO generally operate as part of a joint force. Engineer units may be part of the force actually conducting the operation. In other cases, they may provide support in another country by constructing temporary support facilities for either US forces or the evacuees. Engineer units may also provide topographic products and data tailored to the operation.

Engineers directly participating in the evacuation may perform a variety of tasks, depending on the situation. In a high–threat environment, engineers may perform combat–related engineering tasks. They may construct protective structures for both US forces and evacuees, if necessary. Airfields designated for use in evacuation operations could become damaged—

engineers must be prepared to repair them in case the HN cannot or will not. Engineers may clear landing zones if helicopters will be used either to consolidate evacuees from outlying areas or to move them to waiting ships or other secure areas. If evacuees must be transported on the ground due to a lack of aviation assets or an air defense threat, engineers may be involved in route reconnaissance and mobility–type operations. Field Manuals (FMs) 5–100, 5–101, 5–103, 5–105, and 100–20 and other related doctrinal manuals provide information relating to engineering tasks.

Facility construction may be required to support a NEO. Evacuees may move to another country temporarily before returning to the US. US forces may require a staging area near the country where the NEO will be conducted. If existing facilities are inadequate or not available, engineers may construct a temporary base. The Department of State conducts necessary coordination to obtain clearance to construct and maintain these facilities (early identification of funding sources and authorizations is essential). FMs 5–104, 5–116, 5–166, 31-82, and 100–20 and Technical Manuals (TMs) 5-301-1 through 5-304 provide doctrine relating to facility construction.

RESCUE-AND-RECOVERY OPERATIONS

Rescue-and-recovery operations are sophisticated actions requiring precise execution, especially when conducted in hostile countries. These operations maybe clandestine or overt. They may include the rescue of US personnel or friendly foreign nationals or the location, identification, and recovery of sensitive equipment or items critical to US national security.

Hostile forces can oppose rescue-and-recovery operations. However, these operations may remain unopposed if potentially hostile forces are unaware of them or unable or unwilling to interfere. Stealth, surprise, speed, and the threat of overwhelming US force are some of the means available to overcome opposition. Rescue-and-recovery operations require timely intelligence, including current topographic information. They also require detailed planning, deception, swift execution, and extraordinary security measures. These operations usually involve highly trained special units, but they may also receive support from general purpose forces.

The threat level during rescue and recovery varies depending on the circumstances of the specific operation. Rapid execution of the mission, as well as tight security before and during the operation, will minimize risk. ROE will be established before the operation begins. These rules must be clearly stated and understood by all participants.

During rescue-and–recovery operations, engineer units may provide support to combined arms teams or joint task forces (JTFs). Engineers may provide route reconnaissance, mobility–type operations, construction of helicopter landing zones, and so forth. In these situations, FMs 100–5, and 5–100 and related manuals provide supporting doctrine.

Engineer units may also become involved in constructing staging areas for rescue-and–recovery operations. These areas may include billeting, administrative, and logistical support facilities. The extent of these facilities depends on the projected duration of their occupation.

Construction of staging areas may include mock-ups of buildings, airfields, ports, and so forth to support rehearsals by units participating in the operation.

SHOWS OF FORCE AND DEMONSTRATIONS

US forces deployed abroad lend credibility to US promises and commitments. Shows of force increase the regional influence of the US and demonstrate its resolve to use military force as an element of national power. In addition, the National Command Authorities (NCA) give orders for shows of force or demonstrations to bolster and reassure friends and allies. These operations can influence another government or political–military organization to respect US interests or to enforce international law. These are some examples:

- Forward deployment of military forces.
- Combined training exercises.
- Aircraft and ship visits.
- Introduction or buildup of military forces in a region.

The objective of shows of force and demonstrations must be well-defined and clearly understood. To be effective, the force tasked with the mission must be demonstrably mission-capable and sustainable. The specific requirements for sustainment include adequate command, control, and communications (C^3); intelligence support; interdepartmental and international liaison; and ready and responsive forces. Logistical support for these operations should be based on the possibility that the mission may require the actual use of force.

Political concerns dominate shows of force and demonstrations. Military forces conduct these operations with delicate legal and political constraints. The political will to employ actual force, should a demonstration of it fail, is vital to the success of these operations. Actual combat is not the goal of the operation, but it is a possible outcome. Soldiers involved in these operations must enter them physically and mentally prepared to conduct combat operations. The force coordinates its operations with the country team (or teams, if more than one country is involved). Before commitment, the chain of command must ensure that the members of the force understand the national purpose, ROE, and inherent risks of the operation.

Engineer support to shows of force and demonstrations will normally be conducted in the context of support for a joint or combined force. FMs 5–100 through 5–103 and related manuals provide applicable doctrine for these operations. If a show of force or demonstration continues for a protracted period of time (particularly if it is conducted in an undeveloped area), base-camp, logistical-facility, and LOG construction will become necessary. Under these conditions, manuals such as FMs 5–104, 5–105, and 31-82 and TMs 5-301-1 through 5-304 outline engineer activities required to support the force.

OPERATIONS TO RESTORE ORDER

The US conducts operations to restore order when it is in the national interest to stop a violent conflict in a country or region and to force a return to political and diplomatic methods. (Operations to restore order were previously called "peacemaking.") The US typically undertakes these operations at the request of appropriate national authorities in a foreign nation or to protect US citizens as part of an international, multilateral, or unilateral operation. The long–range goals of an operation to restore order are often unclear. Therefore, these operations are best terminated by prompt withdrawal after a settlement is reached or by rapid transition to a peacekeeping operation (PKO) (see Chapter 6).

The political complexities of operations to restore order require that the available force be sufficient to achieve US objectives and that its use be applied with discretion. ROE tend to be restrictive because the purpose of the force is to maintain law and order. Political considerations may influence force size and composition. The force commander must prepare himself to deal with external pressures not normally associated with military operations. He may have to adjust his operations to reconcile the conflicting demands of political considerations, mission accomplishment, and force protection.

The threat level during an operation to restore order depends on the political and military situation within the area of operations. If one or more of the belligerents in the area oppose the presence of US forces, their capabilities must be considered in the threat assessment. The capabilities of the party that requested US involvement, if one exists, must also be considered. The threat assessment must be continuously updated. It will assist the commander in establishing appropriate steps to ensure adequate force protection.

Engineers involved in operations to restore order support combat operations and perform sustainment and topographic engineering missions. Engineers also support efforts to develop a logistical support base elsewhere in the region, if one is required. Initial force deployment in support of operations to restore order may be conducted by airdrop, airland interdiction, or amphibious operations. To support the buildup of combat forces and permit the flow of logistical support, engineer units may construct new airfields or repair damage to existing ones. Once a forward operating base has been established, expansion of the lodgment area involves engineer support. This may involve constructing protective structures, hardening key facilities, and other activities based on the situation. The duration of the operation, the amount of resistance met by US forces, and the availability of existing facilities will dictate the actual amount and type of engineer support required in the area of operation. FMs 5–100 through 5–103 and related manuals provide applicable doctrine for combat operations. FM 5–104 and related manuals provide guidance relating to sustainment engineering. FM 5–105 provides information on topographic support.

If operations to restore order are conducted in a remote area, it may be necessary to establish a logistical support base or staging area outside the area of operations but still within the region. The Department of State conducts the coordination required to obtain clearance to construct and maintain these facilities. When engineer units support base development, FMs 5–104 and 31-82, TMs 5–301-1 through 5–304, and other related manuals outline engineer activities required to support the force.

STRIKES AND RAIDS

The US executes strikes and raids for specific purposes other than gaining or holding terrain. Strikes and raids can support rescue or recovery operations or destroy or seize equipment or facilities that demonstrably threaten national collective security interests. Strikes and raids can also support counterdrug operations by destroying narcotics production or transshipment facilities or supporting HN activities in this arena. Strikes and raids are the most conventional CONOPs under conditions short of war. The principles of combat operations apply directly (see FMs 5–100, and 100-5 and related manuals). The combatant commander normally plans and executes them.

Before a strike or raid, engineers may construct rehearsal sites for the forces involved in the operation. Topographic engineers may produce large-scale photomaps or graphics to help guide forces to their objectives. Those engineers actually participating in the mission may require refresher training in specialized skills, depending on the specifics of the operation. These skills may include air assault techniques, military operations on urbanized terrain (MOUT), or reorganization as infantry. During strike or raid operations, engineers may be called upon to perform unique missions, in addition to traditional engineer tasks. These missions could include-

- Emplacing and manning roadblocks.
- Moving or disposing of captured equipment.
- Using captured equipment to perform missions.

In some cases, the US may assist with reconstruction in an area where a strike or raid has been conducted. USACE, through an engineer district, can provide teams to assist with damage assessment and contracting for cleanup and repair work. US engineer troop units, if located in the region, may assist by clearing rubble; opening roadways for emergency and medical traffic; and restoring critical facilities, services, and utilities. If USACE teams or engineer troop units are used, they must deploy rapidly to the area to restore key facilities and limit suffering by the local populace.

CORPS CONTINGENCY OPERATIONS

In many CONOPs, an Army corps plans and conducts Army operations as part of a joint or combined force commanded by a Commander in Chief (CINC) (combatant commander) or joint force commander. Sometimes, the corps commander will serve as both Army force (ARFOR) and land component commander in the joint operation. The corps may contribute to a contingency JTF headquarters for campaigns involving all services. A corps that is the largest land force in a theater of operations will often translate strategic goals into tactical objectives. If the corps functions as the nucleus of a JTF headquarters, it is responsible for developing a campaign plan or outline.

To achieve quick, decisive results in CONOPs, the corps must be prepared to deploy and employ its forces simultaneously. In a large operation, while the majority of the force is deploying into the area of operations, lead echelons must lead the way by seizing (if necessary) and securing arrival points for the remaining force. Corps CONOPs will be phased. Phases should begin with planning and preparation and end with contingency force redeployment. However, some phases may not occur at all. The following phases provide the general structure for a CONOPs; they can be adjusted to fit each contingency:

- Predeployment/crisis action.
- Deployment/initial combat operations.
- Force buildup/combat operations.
- Decisive combat operations,
- Redeployment.

For additional information relating to corps contingency planning, see FM 100-15.

Appendix A

Force Protection

PURPOSE

This appendix outlines force-protection measures that engineer units may employ beyond traditional methods, such as triple standard concertina. The intent of this appendix is to provide general force-protection guidance to engineer battalion and brigade commanders and staffs.

GENERAL

This appendix concentrates on the physical-security aspects of force protection. It highlights passive measures for integration into force-protection planning. This appendix also discusses engineer-specific planning considerations required in force protection. It contains very general design criteria. The Security Engineering Manual, published by the Omaha District, United States Army Corps of Engineer (USACE), provides more detailed information on the design of installations and structures requiring protection. This manual also contains information designated "For Official Use Only" on threat capabilities, design of protective measures, and the options outlined in this appendix. Engineer planners should use the USACE manual if they require more specific guidance or if they are designing a permanent facility. Planners should use this appendix in conjunction with Field Manual (FM) 5–103 and Technical Manuals (TMs) 5-301-1/2/3/4 which also contain information applicable to force-protection planning.

ENGINEER PARTICIPATION IN FORCE-PROTECTION PLANNING

Unit operations officers have staff responsibility for their unit's force-protection plan. Engineers will be invoved in force-protection planning from two perspectives. They will prepare their unit's force-protection plans, and they will provide input (and capability) to the unit they are supporting. As with other missions, engineer force-protection planning must be well thought out, logical, and integrated with other staff planning. Force-protection plans or policies must be developed in line with the command estimate process. The engineer must be involved in the intelligence preparation of the battlefield (IPB) process to ensure that engineer intelligence needs are integrated into all reconnaissance and collection plans, information requirements, and priority information requirements.

The engineer must participate in the overall mission analysis, ensuring that engineer mission-essential tasks are integrated into the commander's estimate and that all engineer missions are identified. Using the same analysis process as the maneuver commander, the engineer must focus on—

- Engineer specified and implied tasks.
- Assets available to enhance force protection.
- Conatraints.
- Restrictions.
- Risks.
- Time analysis (the Security Engineering Manual calls this "response time").
- Essential tasks.

In developing a force-protection policy, the following process should be used:

- Determine the composition of assets (personnel, equipment, and facilities.
- Define the threat and attack probability.
- Determine levels of protection for each asset.
- Identify constraints.
- Design protective systems to counter threats.

The engineer must ensure that the maneuver staff and commander develop a force-protection policy based on the threat. The plan must balance the attack probability, the consequences of inadequate protection, and the cost of adequate protection (risk level). The commander must set the priority of protection for United States (US) forces and equipment, local assets, infrastructure, and the local populace.

THE THREAT

Engineers must understand threat components, which they must then counter. The threat has three components: aggressors; their tactics; and their associated weapons, explosives, and tools. There are four types of aggressors that engineers must understand and plan against in a low intensity conflict (LIC) environment:

- Criminals. Subdivided into three cateogories: unsophisticated, sophisiticated, and organized. Unsophisticated criminals are unskilled in the use of weapons and tools and have no formal organization. Their targets are those that meet their immediate needs, such as drugs, money, and pilferable items. They are interested in targets that pose little risk. Sophisticated criminals working singly are organized and efficient in the use of certain weapons and tools. They target high–value assets and frequently steal large quantities. Organized criminal groups are sophisticated and rely on specialists to obtain equipment to achieve specific goals. Targets of organized criminal groups may involve large quantities of money, equipment, arms, ammunition, and explosives.
- Protestors. Categorized as either vandals, activists, or extremists. Engineers must be concerned with all violent protectors. Protectors are politically or issue-orientated and act out of frustration, discontent, or anger. Their primary objectives include destruction and publicity. Vandals and activists are unsophisticated and superficially destructive. They generally do not intend to injure people. Extremist groups are moderately sophisticated and more destructive. Their actions are frequently overt and may involve individuals as targets.
- Terrorists. Oriented on an ideology, a political cause, or an issue. Terrorists commonly work in small, well-organized groups. They are sophisticated and possess an efficient

planning capability. Terrorist objectives include death, destruction, theft, and publicity. Terrorist groups are generally classified by their government affiliation. They are categorized as nonstate-supported, state-supported, or state-directed. Chapter 1 contains a detailed description of these categories. Chapter 5 lists specific terrorist tactics.

- Subversives. Classified into two groups: saboteurs and spies. Saboteurs include guerrillas and commandos. They are very sophisticated and highly skilled and employ meticulous planning. Saboteurs commonly operate in small groups and have an unlimited arsenal. Their objectives include death and destruction. They often target mission-critical personnel, equipment, or operations. Spies are highly skilled and very sophisticated. They are generally foreign agents but frequently employ insiders. They target military information and attempt to avoid detection. In some cases, they may use the activities of other aggressors.

Threat personnel employ a wide range of tactics to accomplish their objectives. These strategies have been categorized into 15 tactics, which are specific methods of achieving an aggressor's goals. Not included in this appendix are other tactics that engineers have little capability of countering. These include kidnapping, hijacking, and assassinations that are executed away from a facility or installation. The following descriptions of aggressor tactics will assist engineer planners in developing protective methods, devices, facilities, and systems:

- Moving-Vehicle Bomb. Used when an aggressor's goal is to damage or destroy a facility (or assets within a facility) or to kill people within the blast area. The moving-vehicle bomb is a suicide attack where an explosive-laden vehicle is driven into a facility and detonated.
- Stationary–Vehicle Bomb. Used when an aggressor's primary objective is to damage or destroy a facility (or assets within a facility). This type of bomb maybe detonated by time delay or remote control. This attack has three versions:

  - An explosive-laden vehicle is driven to a preselected location and abandoned.
  - Explosives are placed in an unsuspecting person's car. He then unknowingly delivers the bomb to the targeted facility.
  - Someone is coerced into delivering a vehicle bomb.

- Exterior Attack. Used when an aggressor's goal is to damage or destroy a facility (or assets within a facility) and kill or injure its occupants. This attack is at close range of a facility or exposed asset. Using clubs, rocks, improvised incendiary devices, hand grenades, or hand–placed bombs, the aggressor attempts to inflict destruction and death.
- Standoff Weapons Attack. Used when an aggressor's goal is to damage or destroy a facility (or assets within a facility) and kill or injure its occupants. These attacks are executed using military or improvised direct-and indirect-fire weapons, such as antitank weapons and mortars.
- Ballistic Attack. Used when an aggressor's goal is to kill or injure a facility's occupants. Using small arms at varying distances, the aggressor attempts to inflict death.
- Forced Entry. Used when an aggressor's goals are to steal or destroy assets, compromise information, or disrupt operations. Using small arms or forced-entry tools, the aggressor enters a facility through an existing passage or creates a new opening in the facility.
- Covert Entry. Used when an aggressor's goals are identical to those listed for the forced-entry tactic. The difference in these entries is that the aggressor will attempt to enter the facility covertly using false credentials. The aggressor may attempt to carry weapons or explosives into the facility.

- Insider Compromise. Used when an aggressor's goals are similar to those listed for the forced-entry tactic. The aggressor uses an insider (one who has legitimate access to a facility) to accomplish their prescribed objectives.
- Electronic Eavesdropping. Used by an aggressor to monitor electronic emanations from computers, communications, and related equipment. This eavesdropping is normally done from outside a facility or restricted area.
- Acoustical Eavesdropping. Used by an aggressor (using a listening device) to monitor voice communication and other audible information.
- Visual Surveillance. Used by aggressors employing ocular and photographic devices to monitor facility, installation, and mission operations.
- Mail Bombs. Used when the aggressor's objective is to kill or injure people. Small bombs or incendiary devices are incorporated into envelopes or packages that are delivered to the targeted individual.
- Supplies Bombs. Used when the aggressor's objective is to kill or injure people or destroy facilities. Bombs or incendiary devices, generally larger than those found in mail bombs, are incorporated into various containers and delivered to facilities or installations.
- Airborne Contamination. Used when the aggressor's objective is to kill people, The aggressor uses chemical or biological agents to contaminate the air supply of a facility or installation.
- Waterborne Contamination. Used when an aggressor's objective is to kill people. The aggressor uses chemical, biological, or radiological agents to contaminate the water supply of a facility or installation.

Aggressors use various types of weapons, explosives, and tools to attain their objectives. Weapons range from clubs and rocks to mortars. Explosives are commonly used to destroy facilities and housing assets and to kill people. Tools are primarily used in forced-entry operations to breach protective components or barriers. Understanding the aggressor's options will aid the engineer in protecting forces from these items. Listed below are various weapons, explosives, and tools and their potential use:

- Rocks and Clubs. Used in exterior building attacks to damage exterior building components or exposed assets or to injure people.
- Incendiary Devices. Used to damage the facility's exterior or sabotage other assets. These include hand–held torches and improvised incendiary devices (IID). An example is a "Molotov cocktail."
- Firearms. Used in the ballistic tactic to attack facility assets from a distance and in the forced-entry tactic to overpower guards. These include pistols, rifles, shotguns, and submachine guns, both military and civilian. Weapons capabilities are outlined in the
- Security Engineering Manual.

  Antitank Weapons and Mortars. Used in standoff attacks of facilities. For example, the direct-fire antitank weapons most often used by terrorists are the Soviet, rocket-propelled grenade RPG–7 and the US light antitank weapon (LAW). These weapons increase the terrorist's ability to penetrate and damage a facility and to kill or injure people. Mortars are indirect–fire weapons and include both military and improvised versions.
- Nuclear, Biological, and Chemical. Delivered as airborne or waterborne gases, liquids, aerosols, or solids. Very powerful chemical agents can be manufactured with relative ease from commercially available products. Biological agents can be grown in unsophisticated

home laboratories. Radiological agents are radioactive elements that pose a potential threat to water supplies. They can be delivered in liquid or solid form.

- Improvised Explosive Devices (IED). Used in the exterior–attack, mail– and supplies–bomb-delivery, forced-entry, covert-entry, and insider-compromise tactics to destroy assets and to injure or kill people. They are commonly "homemade" bombs made of plastic explosives or trinitrotoluene (TNT). Plastic explosives are chosen by terrorist and extremist protester groups because they are easily molded, stable, and difficult to detect.
- Hand Grenades. Used in exterior attacks to injure or kill people. These include common military antipersonnel and fragmentation hand grenades.
- Vehicle Bombs. Used to destroy facilities and kill people. They contain large quantities of explosives and have the potential to do catastrophic damage.

Potential aggressors have access to a wide variety of tools, ranging from forced-entry tools, (hand and power tools, cutting torches, and burn bars), to sophisticated surveillance tools and devices. The quality and effectiveness of tools and devices used depends on the type of aggressor. The more sophisticated, trained, and organized the aggressor is, the more dangerous his tools and devices will be.

PROTECTIVE MEASURES AND TECHNIQUES

Engineers may play a significant role in protecting deployed US forces. They have the capability, when given time, priority, and a thorough IPB, to effectively establish defensive measures to protect forces, facilities, and equipment from potential aggressors. The following list of measures will enhance a force's survivability. The specific options that the engineer planner selects will be based on the-

- Specific threat in the area of operations.
- Degree of protection required.
- Time available.
- Materials available.

To tailor the force-protection package to the local threat, the engineer planner should consult the provost marshal and intelligence personnel when selecting options from these lists.

Minimum–Measures Checklist.

Basic considerations—

- ❑ Eliminate potential hiding places near facilities.
- ❑ Provide an unobstructed view around all facilities.
- ❑ Site facilities within view of other occupied facilities.
- ❑ Locate assets stored on–site but outside facilities within view of occupied rooms of the facilities.
- ❑ Minimize the need for signs or other indications of asset locations.
- ❑ Minimize exterior signs that may indicate location of assets.
- ❑ Provide a 170–foot minimum facility separation from installation boundaries.

- ❑ Eliminate lines of approach perpendicular to buildings.
- ❑ Minimize vehicle and personnel access points.
- ❑ Eliminate parking beneath facilities.
- ❑ Locate parking as far from facilities as practical.
- ❑ Illuminate building exteriors or exterior sites where assets are located.
- ❑ Secure access to power and/or heat plants, gas mains, water supplies, and electrical service.
- ❑ Locate public parking areas within view of occupied rooms or facilities.
- ❑ Locate construction staging areas away from asset locations.
- ❑ Locate facilities away from natural or man–made vantage points.
- ❑ Locate facilities' critical assets within areas that do not have exterior walls, when possible.
- ❑ Minimize window area.
- ❑ Cover windows next to doors so that aggressors cannot unlock the doors through them.
- ❑ Secure exposed exterior ladders and fire escapes.
- ❑ Design building layout so that there are no areas hidden from view from control points or occupied spaces.
- ❑ Arrange building interiors to eliminate hiding places.
- ❑ Locate assets in spaces occupied 24 hours a day, when possible.
- ❑ Locate activities with large visitor populations away from protected assets when possible.
- ❑ Locate protected assets in controlled areas where they are visible to more than one person.
- ❑ Place mail rooms on the perimeter of facilities.
- ❑ Provide emergency backup power generation for critical activities/facilities.

Moving-and Stationary-Vehicle-Bomb Checklist. (Note: When designing/planning to defeat the moving-vehicle tactic, you must plan to defeat the vehicle, not the driver. For example, a guard with an M16 will not stop a vehicle.) In addition to the minimum–level measures, consider the following measures to counter a moving– or stationary-vehicle bomb threat:

Site design—

- ❑ Provide exclusive standoff zones for facilities.
- ❑ Provide nonexclusive standoff zones for facilities (high to very high threat).
- ❑ Cluster facilities in common exclusive and nonexclusive standoff zones when possible.
- ❑ Use passive vehicle barriers, Table A-1 shows the maximum speed that a vehicle of a given weight can impact a barrier and still be stopped by that type of barrie.,

Table A-1. Passive vehicle-barrier capabilities.

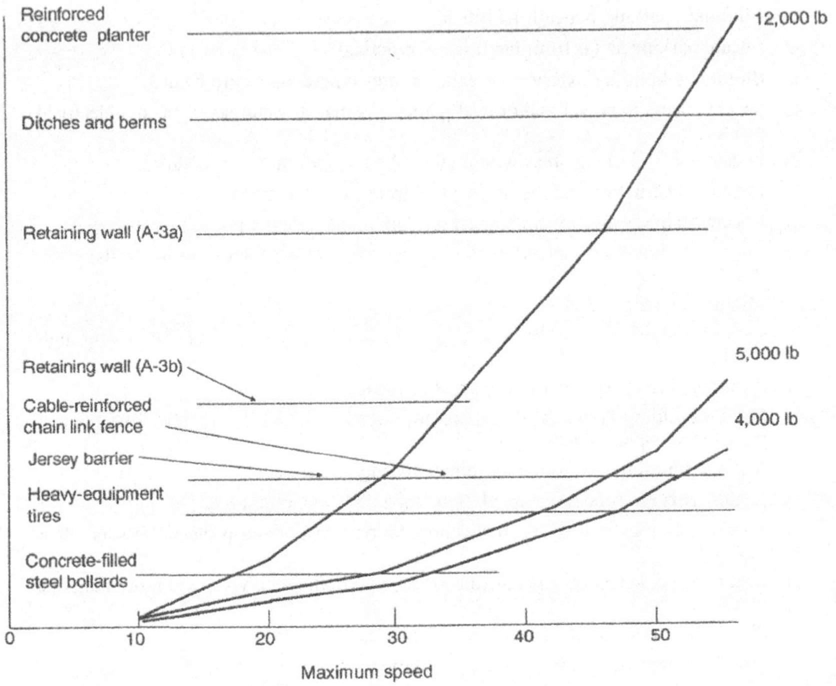

- ❑ 8– to 12–inch straight curbs.
- ❑ Chain link or ornamental fence.
- ❑ Decorative posts (4-inch diameter at 4-foot centers) with or without chain.
- ❑ Trees and shrubs.
- ❑ Shallow ditches.
- ❑ Low berms.
- ❑ Half–buried heavy-equipment tires (see Figure A–l).
- ❑ Concrete barriers (Jersey barrier) (see Figure A-2, page A-8).
- ❑ Concrete retaining walls (see Figures A–3a and A–3b, page A–9).
- ❑ Reinforced concrete planters (see Figure A–4, page A–10).
- ❑ Concrete-filled steel bollards (see Figure A–5, page A–10).
- ❑ Cable-reinforced chain link fences. (Chain link fence reinforced with two 3/4-inch–diameter cables, one at 30 inches and one at 35 inches above ground level.)

- ❑ Ditches (triangular or trapezoidal) (see Figures A-6a and A-6b, page A-11).
- ❑ Berms (see Figure A-7, page A-11).
- ❑ Sandbags.
- ❑ Tetrahedrons (steel or concrete).
- ❑ Concrete-, rock-, or sand-filled 55-gallon drums connected with wire rope.

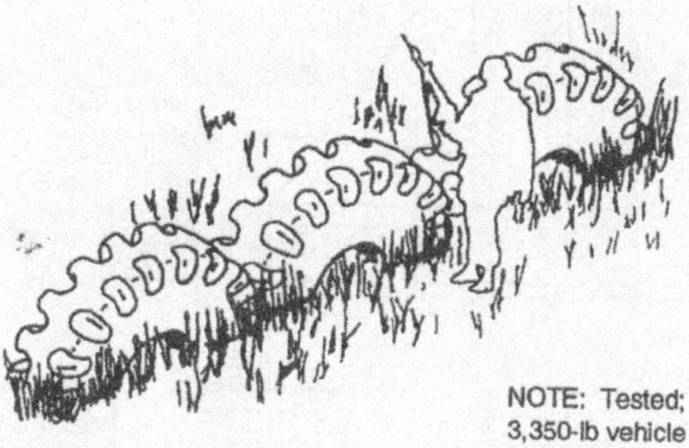

NOTE: Tested; stopped a 3,350-lb vehicle traveling at 50 mph.

Figure A–1. Heavy–equipment tires.

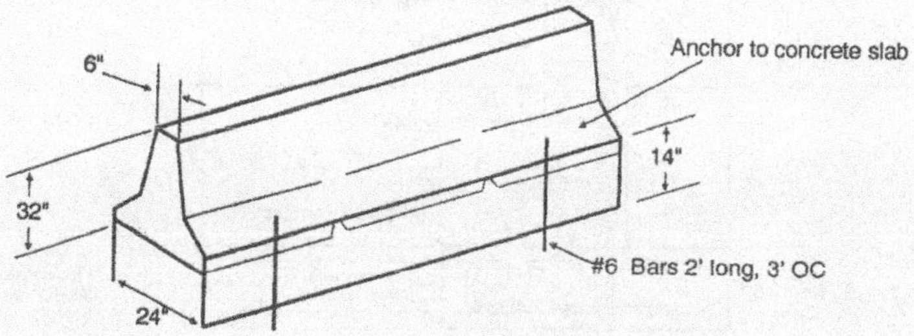

NOTE: Tested; stopped a 4,000-lb vehicle traveling at 50 mph.

Figure A–2. Jersey barrier.

NOTE: Tested; stopped a 15,000-lb vehicle traveling at 40 mph.

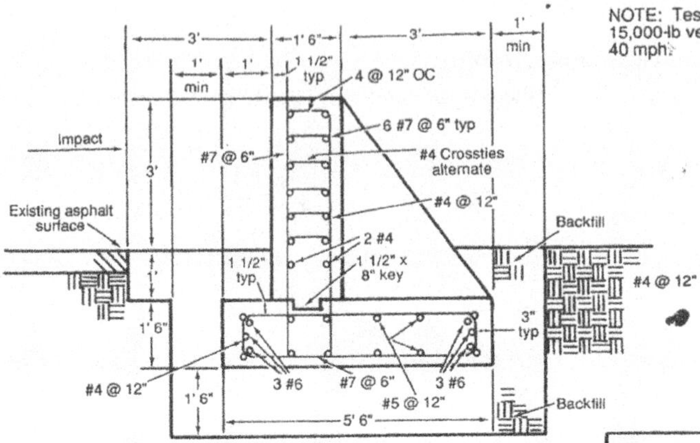

A-3a. Retaining wall

Each wall must be calculated based on soil conditions at that site.

Concrete; $f'_c = 3{,}000$ *psi*

Reinforcement steel bars; $f_y = 60$ *ksi*

1 1/2" concrete cover all around except as noted.

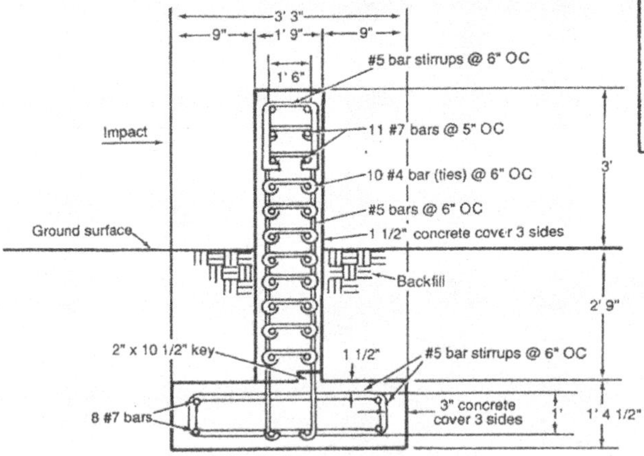

NOTE: Tested; stopped a 15,000-lb vehicle traveling at 30 mph.

A-3b. Retaining wall

Figure A-3. Concrete retaining walls.

NOTE: Tested; stopped a 15,000-lb vehicle traveling at 50 mph.

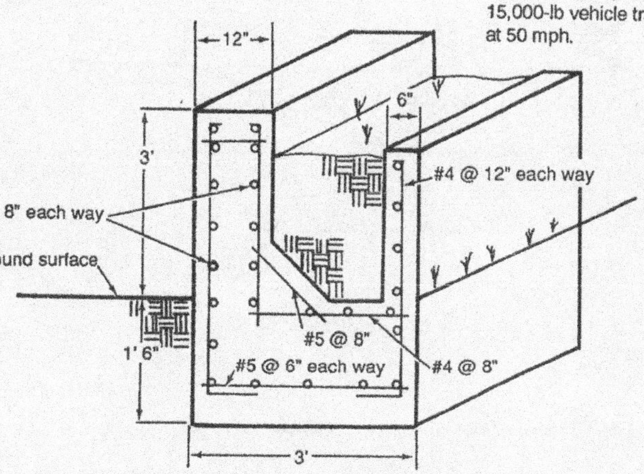

Figure A–4. Reinforced concrete planter.

NOTE: Tested; stopped a 4,500-lb vehicle traveling at 30 mph.

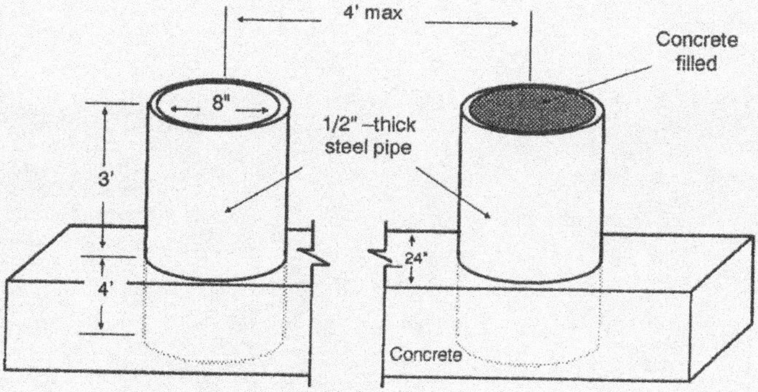

Figure A–5. Concrete–filled steel bollards.

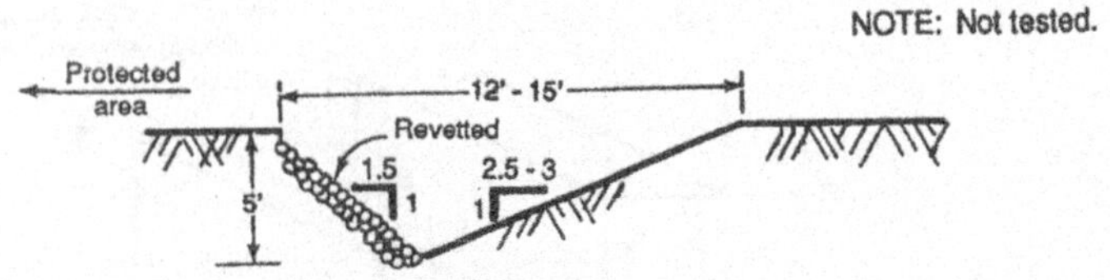

Figure A-6a. Triangular ditch.

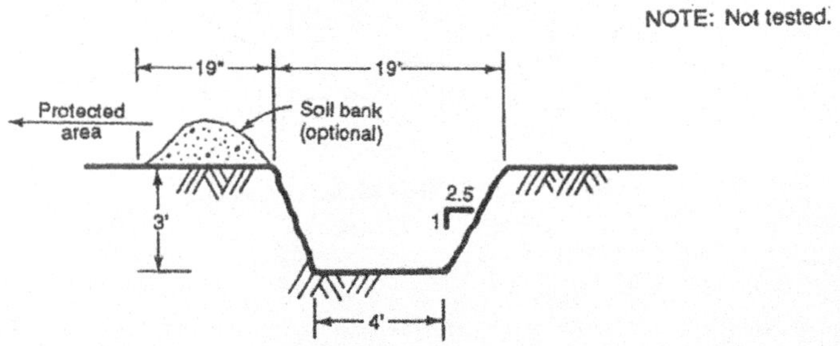

Figure A-6b. Trapezoidal ditch.

Figure A–6. Ditches.

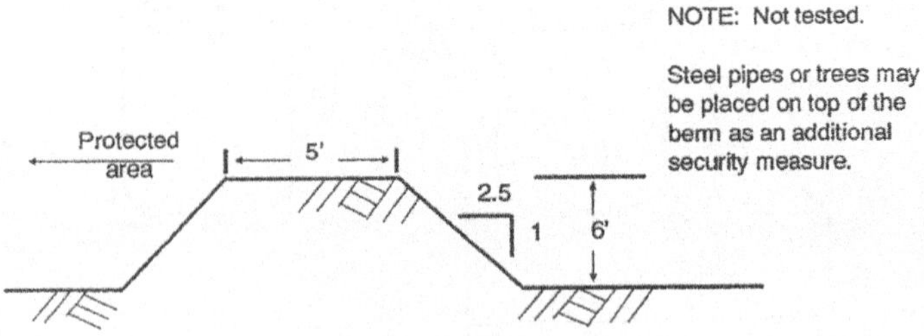

Figure A–7. Berm.

❑ Use active vehicle barriers (see Table A-2)—

- ❑ Drum-type barriers (see Figure A-8).
- ❑ Plate-type barriers.
- ❑ Portable barriers (see Figure A-9).
- ❑ Crash beams.
- ❑ Crash gates (sliding gate) (see Figure A–10, page A-14).
- ❑ Cable-beam barrier (see Figure A-11, page A-14).
- ❑ Retractable bollards (see Figure A-12, page A-15.).

Table A–2. Active vehicle barriers capabilities.

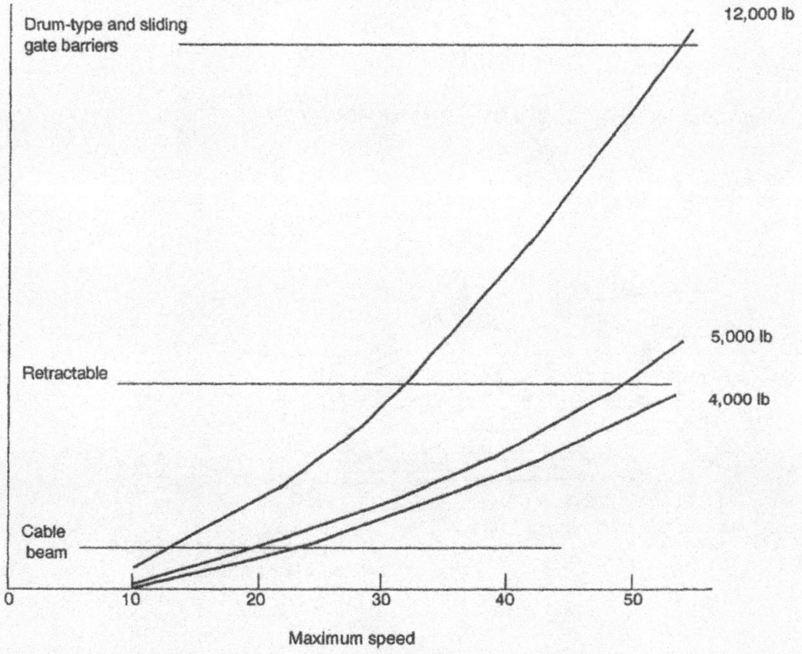

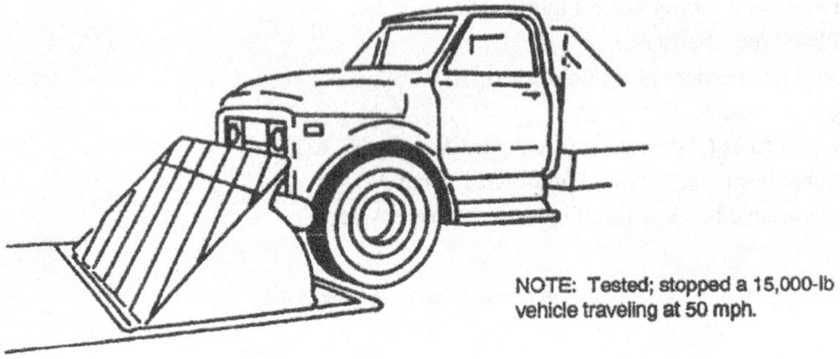

Figure A–8. Drum–type barrier.

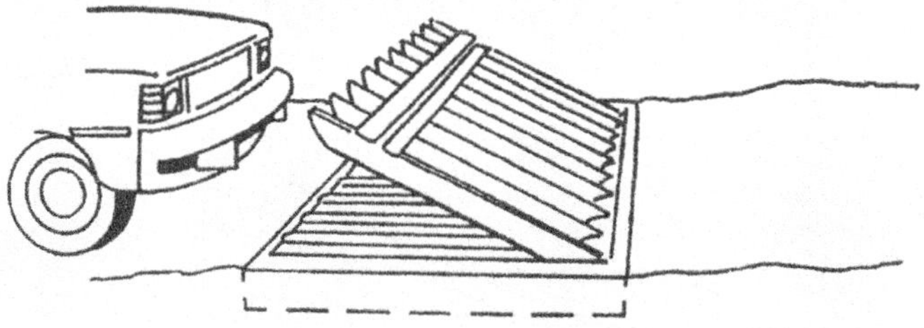

Figure A–9. Portable barriers.

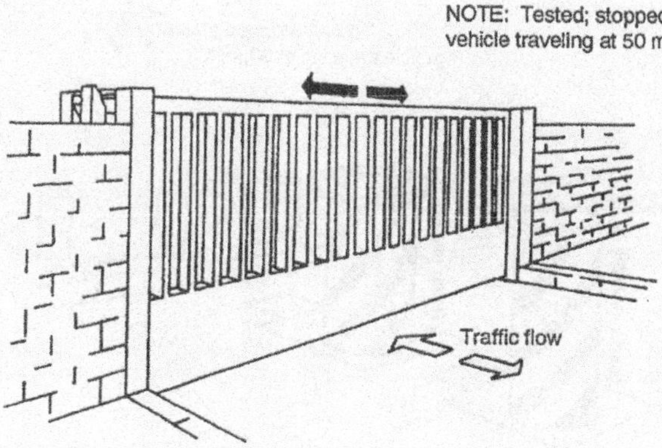

Figure A–10. Sliding gate.

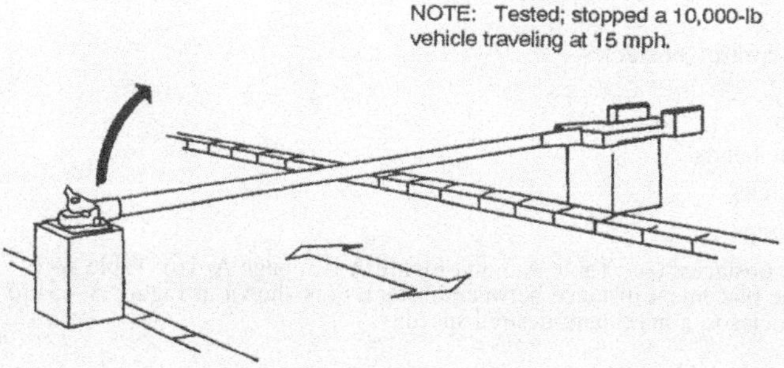

Figure A–11. Cable–beam barrier.

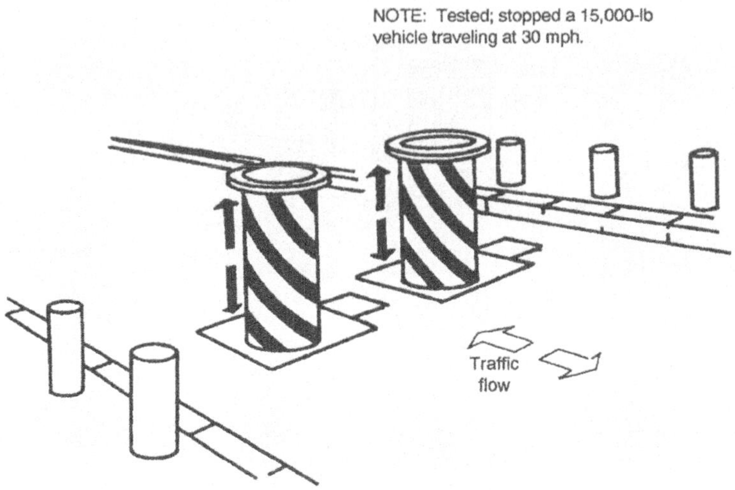

Figure A–12. Retractable bollards.

- Use speed-control obstacles–
  - S-curves.
  - 90-degree bends.
  - Traffic circles.
  - Speed bumps.
  - Concrete obstacles (see Table A-3 and Figure A–13, page A–16). Table A–3 shows the placement distance between obstacles (as shown in Figure A–13) to limit vehicles to a maximum desired speed.

Table A–3. Obstacle spacing versus desired maximum speed.

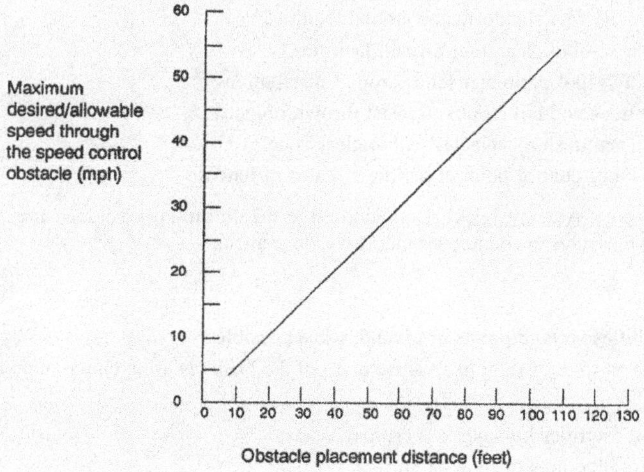

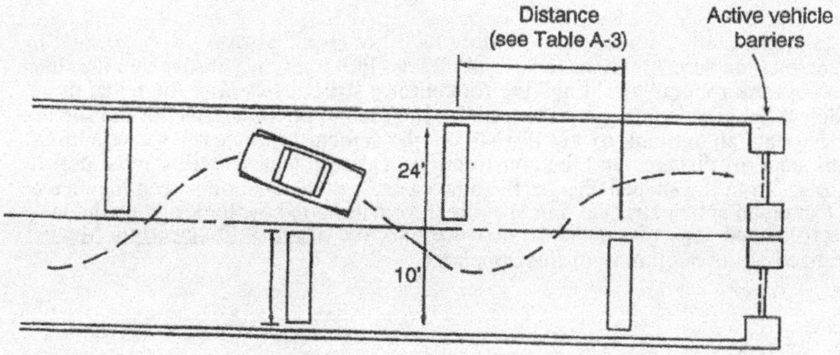

Figure A–13. Concrete–obstacle placement.

Exterior-Attack Checklist. In addition to the minimum-level measures, consider the following measures to counter an exterior attack of a facility or installation:

Site design—

- ❑ Provide a 50–foot standoff zone around facilities.
- ❑ Provide a 30–foot clear zone around facilities.
- ❑ Provide a 7–foot perimeter fence around installations.
- ❑ Provide trees and tall shrubs to resist thrown objects.
- ❑ Do not place trash receptacles within clear zone.
- ❑ Provide entry control point at perimeter gates in fences.

Standoff-Weapons-Attack Checklist. In addition to the minimum-level measures, consider the following measures to counter a standoff weapons attack:

Site design—

- ❑ Site facilities on high points of ground, when possible.
- ❑ Block direct lines of sight to sensitive areas of the facilities using one or more of the following:
  - ❑ Other facilities housing less critical assets.
  - ❑ Dense plantings of trees or shrubs.
  - ❑ Obscuration fences.
  - ❑ Walls.
  - ❑ Earth berms.
- ❑ Provide predetonation screens surrounding facilities (see Figure A-14, page A-18). A predetonation screen may have several effects. Preferably it will damage the fuse on an antitank rocket, "dudding" the rocket (any structure behind the fence must still defeat the kinetic energy of the round). The second possible effect of the screen is to detonate an antitank rocket if it strikes the screen. In this case, the combination of standoff distance and the construction of the protected facility must defeat the gas jet from the shaped charge. In some cases, a predetonation screen may have no effect on an antitank rocket. For specific information on how to design predetonation screens and determine stand off distances, see the Security Engineerting Manual referenced at the beginning of this appendix.

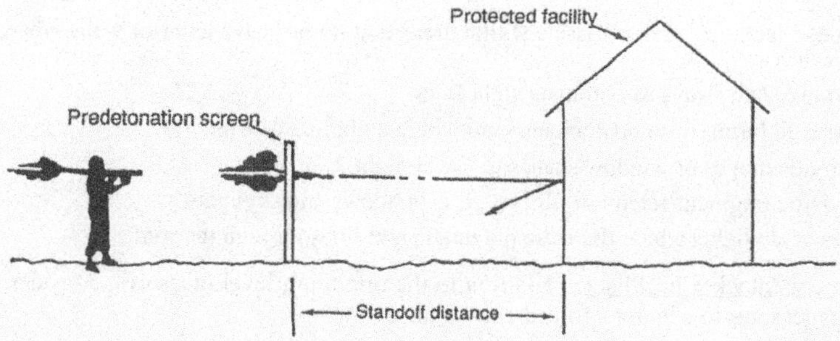

Figure A–14. Predetonation Screen.

Building utilization—

- ❑ House critical assets away from the exterior of facilities.
- ❑ Provide an 8–foot-wide area (minimum) within the facilities (around the facilities' perimeter) that houses only noncritical assets.
- ❑ Place assets in protective areas below grade or underground.
- ❑ Provide entry foyers.
- ❑ Arrange furniture so that people are out of sight of windows.
- ❑ Minimize exterior access by blocking some of the existing doors.

Ballistic-Attack Checklist. In addition to the minimum-level measures, consider the following measures to counter a ballistic attack:

Site design—

- ❑ Site facilities on high points of ground, when possible.
- ❑ Block direct sight line to sensitive areas of the facilities using one or more of the following:
    - ❑ Other facilities housing less critical assets.
    - ❑ Dense plantings of trees or shrubs.
    - ❑ Obscuration fences.
    - ❑ Walls.
    - ❑ Earth berms.

Building utilization—

- ❑ House facilities' critical assets within areas that do not have exterior walls when possible.
- ❑ Arrange entryways to eliminate sight lines.
- ❑ Arrange furniture so no personnel are within sight line of windows.
- ❑ Provide drapes or window shades to use at night.
- ❑ Provide fragment-retention film on glass windows and skylights.
- ❑ Cover skylights where there are potential sight lines through the roof.

Forced-Entry-Attack Checklist. In addition to the minimum level measures, consider the following measures to counter a forced-entry attack:

Site design—

- ❑ Provide unobstructed view around facilities.
- ❑ Site facilities away from installation boundaries and areas open to the public.
- ❑ Site facilities within view of other occupied facilities.
- ❑ Locate assets stored on–site but outside the facilities within view of occupied rooms in the facilities.
- ❑ Use screens or vegetation to conceal assets located on-site but outside of any facilities.
- ❑ Denote boundaries with fences or freestanding walls.
- ❑ Conceal and lock access ports to utilities.
- ❑ Illuminate building exterior or exterior sites where assets are located.
- ❑ Minimize shrubbery or other landscaping higher than 1 foot above finished floor height of buildings.

Building utilization—

- ❑ Locate protected assets as far from exterior walls as possible.
- ❑ Cluster assets to minimize the number of hardened spaces within facilities, or
- ❑ Distribute assets in large facilities to increase the difficulty of access, or
- ❑ Locate assets in basement spaces or underground facilities.
- ❑ Plan alternate locations for assets to allow the change of location periodically or during higher threat conditions.
- ❑ Locate assets above, below, or oblique to sight lines.
- ❑ Design building layout so that there are no areas hidden from view from control points or occupied spaces.
- ❑ Arrange building interiors to eliminate hiding places.
- ❑ Locate assets in spaces occupied 24 hours a day, when possible.
- ❑ Provide temporary sites for assets that may be threatened during higher threat conditions in spaces occupied 24 hours a day.

- ❑ Lay out buildings to conceal assets and make access difficult for intruders.
- ❑ Do not expose a buildings' structure or otherwise advertise the nature of the facilities' construction.
- ❑ Secure skylights.
- ❑ Secure roof access hatches.
- ❑ Eliminate extierior access to roofs, where possible.
- ❑ Secure exposed exterior ladders and fire escapes.

Cover-Entry-and Insider-Compromise-Attack Checklist. In addition to the minimum-level measures, consider the following measures to counter covert entry and insider aggression. This list does not include the many options available to detect unauthorized activity or to control access.

- ❑ Locate activities with large visitor populations away from protected assets.
- ❑ Locate protected assets in common areas where they are visible to more than one person.
- ❑ Minimize the number of entrances into protected areas.
- ❑ Place protected assets in restricted areas to which access is limited to authorized personnel or escorted visitors.
- ❑ Compartmentalize protected assets within restricted areas to limit access among authorized personnel.
- ❑ Establish access control points outside of controlled areas to search personnel and carry–in items.
- ❑ Provide metal or metal-clad doors and metal frames on exterior walls of access control points.
- ❑ Provide blast-resistant doors between access control points and protected areas.

Electronic-and Acoustical-Eavesdropping Checklist. In addition to the minimum-level measures, consider the following measures to counter eavesdropping:

Site design—

- ❑ Eliminate hiding places near the facilities from which aggressors can eavesdrop.
- ❑ Control access to exterior areas around the building.
- ❑ Control access to utility openings.
- ❑ Locate protected assets in buildings away from installation boundaries.

Building utilization—

- ❑ Locate protected assets away from the exterior of buildings and away from interior, uncontrolled areas.
- ❑ Surround protected areas with controlled–access rooms that house noncritical assets.
- ❑ Do not allow operable windows in protected areas.

- ❑ Account for all telephone and other cables.
- ❑ Disconnect and remove unused cables.
- ❑ Do not use intercom systems.

Visual-Surveillance Checklist. In addtion to theminimum-level measures, consider the following measures to counter visual surveillance:

Site design—

- ❑ Site facilities on a high point of land, when possible.
- ❑ Obscure assets from sight lines; use one or more the following methods:
  - ❑ Screen critical assets with facilities housing less critical assets.
  - ❑ Use visual barriers to block views of the asset (dense plantings of trees or shrubs, decorative wal1s or fences, earth berms, or obscuration fences.

Mail-Bomb-Tactic Checklist. In addtion to the minimum-level measures, consider the the following measures to counter a mail–bomb threat (measures for detecting bombs are not included):

Building utilization—

- ❑ Locate mail rooms away from main facility entrances, areas of high density population, and critical assets.
- ❑ Provide a bomb-disposal container in the mail room.
- ❑ Place mail rooms on the perimeter of facilities, or
- ❑ Use a central mail–handling facility that is separate from protected facilities.

Supply–Bomb–Tactic Checklist. In addition to the minimum–level measures, consider the following measures to counter a supply-bomb threat (measures for detecting bombs or explosives are not included):

Site design—

- ❑ Park vehicles away from loading dock areas.
- ❑ Site facilities so that loading dock areas are away from other facilities or high density population areas.
- ❑ Site facilities as far from other facilities as possible.

Building utilization—

- ❑ Place supplies so they are at least 8 feet from facility walls to provide a standoff distance between bombs and walls.
- ❑ Place receiving areas away from facility entrances.
- ❑ Locate assets away from receiving areas.
- ❑ Provide a separate facility for handling and receiving supplies.

Appendix B

Deployment Considerations

PURPOSE

This appendix provides planning considerations for the independent deployment of engineer units. During these deployments, units are generally augmented with combat support and combat service support assets. An engineer headquarters usually provides command and control for the deploying task force. The headquarters is responsible for planning the deployment for all assigned and attached units.

GENERAL

Each engineer unit deployment is slightly different. The considerations in this appendix are general in nature and may not apply in all cases. The intent of this appendix is to provide a list of issues that the engineer leader should consider when planning a deployment. Situation–dependant conditions (such as threat level) or local policies may dictate modification of some of the issues.

DEPLOYMENT PLANNING

The following checklists provide engineer commanders and staff officers with a variety of issues that they should consider when planning a deployment. The checklists are broken down by general functional area to assist in assigning responsibilities to various staff sections. As a unit gains experience conducting deployments, it should develop standard operating procedures (SOPs) and expand these checklists based on lessons learned.

General-Planning-Considerations Checklist. The following are general considerations to use when planning a deployment:

- ❑ Location of the exercise.
- ❑ Forces participating in the exercise (Army, Air Force, Navy and/or Marines, host nation (HN), and other nations).
- ❑ Memorandums of Understanding (MOUs) between the HN and the United States (US).
- ❑ Specified missions.

- Implied missions. Implied missions may include deployment of the task force, base-camp construction, maintaining communications, logistical sustainment, cooperation with HN forces, and redeployment.
- Points of contact (POCs). A readily available file of names, telephone numbers, and addresses should be maintained by the Operations and Training Officer (S3), updated regularly, and disseminated to all planners concerned. This list should include the US military attache and the chief of the security assistance organization (SAO).
- Command relationships for the exercise. A clear chain of command must be established, to include determining whether participants (Active Component (AC) and Reserve Components (RC)) are assigned or attached to the task force.
- Appropriate headquarters element required to command and control the operation. If an engineer battalion augmented with combat support and combat service support assets conducts an exercise, a group or brigade headquarters element may be appropriate. This headquarters will provide command and control, act as an intermediary with the next higher headquarters, and coordinate visits by very important persons (VIPs) and the media.
- Unit in–progress reviews. For meetings where agreements on services, contracts, construction plans, and so forth, are being made, ensure that key battalion staff or duration task force oversea deployment training (ODT) personnel are involved. Ensure that these personnel have adequate documentation on all agreements reached. Ensure that the US country team is provided with adequate information on what aspects of the exercise are easy to change and what portions are not. This is especially important when negotiating agreements with HN representatives. If possible, place a task force representative on the US planning and negotiating team when exercise plans are being finalized before the start of the exercise (STARTEX).
- In–progress reviews to the US country team and higher echelons.
- RC constraints. AC units must attune themselves to the constraints that RC units have when they are part of a task force. RC units have a compressed time line for preparing for exercises and must be notified of the specific missions, exercise dates, transportation modes, and so forth, as early as possible.

Operational-Considerations Checklist. The following considerations apply to the operational aspects of a deployment:

- Training objectives and skill levels required of US forces. When dealing with RC units, pay particular attention to tasks ordinarily performed by full-time technicians and Active Guard Reserve (AGR) personnel, such as daily equipment maintenance and supply requisitions.
- Exercise objectives and corresponding skill levels needed by HN forces.
- Psychological and civic action exercise objectives. Ensure that these objectives are coordinated with the theater Commander in Chiefs (CINC) staff and the country team.
- Training restrictions. Determine restrictions on the use of pyrotechnics and live ammunition. Determine any limits on the size of unit training exercises. Determine procedures and coordination requirements for shutting off power to telephone/power lines on drop zones (DZs) for airborne/air assault training operations.

- ❑ Exercise phasing. Phasing will be affected by project requirements, the necessity to establish a base camp, the amount of survey work required, and so forth.
- ❑ Milestones. Milestones will be determined by exercise phasing, critical events, and external constraints.
- ❑ Personnel requirements. These are based on mission requirements and unit capabilities. Nondeployable personnel must be identified. Critical skills must be designated and filled.
- ❑ Equipment requirements. These are based on mission requirements and unit capabilities. Specialized units or equipment must be requested.
- ❑ Other units in the area that may provide assistance. Special Forces (SF), civil affairs (CA), and psychological operations (PSYOP) units have a great deal of capability relating to local laws, customs, and language. Engineer units should request augmentation by CA personnel for their language capability and understanding of the local populace.
- ❑ Rules of Engagement (ROE).
  - ❑ Ensure that ROE concerning the use of deadly force and other self–protection measures are clearly defined and widely disseminated
  - ❑ Ensure that policies on who will carry arms (loaded or unloaded) and the types and level of tactical training have been coordinated with the country team and approved by the US ambassador.
  - ❑ Establish procedures for alert and recall under various threat conditions. Determine what communications systems and procedures will be used if radios become unusable.
- ❑ Support requirements. These requirements include communications assets; water purification and storage; shower and laundry facilities; mess operations; health medical support; bulk petroleum, oils, and lubricants (POL) storage and dispensing direct support maintenance; transportation augmentation; and aviation support.
- ❑ Reproduction requirements for classified and unclassified material. Procure assets to meet needs that exceed the current unit capability.
- ❑ Compatibility of various computer systems. Address compatibility within the task force and between the task force and the theater CINC, the continental United States (CONUS) support units, and in-country support units.
- ❑ Coordination with augmentation units. If deployment for training (DFT) augmentation is required to accomplish the mission, unit types and desired capabilities should be defined early and communicated to higher headquarters for unit assignment.
- ❑ Organization of rear detachment. A rear-detachment commander must be identified. Procedures and criteria for a rear detachment must be developed.
- ❑ Effective dates of attachment of supporting units. Effective dates will be determined by exercise phasing and the requirements for having units on-site. The dates also depend on whether units will be attached before departure from their home station or at the point of embarkation or debarkation. Incorporate logistics planners early in the planning process. They should develop a logistics intelligence file on available resources in the area of operation. This information may be used to minimize transportation requirements.

- ❑ Composition of predeployment site-survey team. Consider for inclusion:
  - ❑ Designated task force commander or his deputy.
  - ❑ Battalion construction officer who will prepare designs and bills of materials.
  - ❑ Noncommissioned officers (NCOs) with experience in the expected type of construction.
  - ❑ Surveyors and draftsmen with equipment.
  - ❑ Logistics-operations officers.
  - ❑ Two bilingual soldiers with capabilities in the HN language-ideally one officer and one noncommissioned officer.
  - ❑ HN representative as determined by the US embassy.
  - ❑ SAO representative.
- ❑ Mission of predeployment site-survey team. Ensure that the predeployment survey team addresses transportation, linguistic support, billeting, and messing requirements. The team should obtain photographs of the operational area (if permitted by the HN), particularly maneuver areas, bivouac sites, and airfields. Operational considerations that the survey team should address include-
  - ❑ Location for the base camp or bivouac sites.
  - ❑ Location of potable-water supplies.
  - ❑ Location of nonpotable water and requirements for purification.
  - ❑ Distance from the deployment site to logistical support, such as rations, repair parts, and medical support.
  - ❑ Availability of local logistical support. Research local purchase or contracting of repair parts, ration supplements, POL, transportation assets (for personnel and equipment), services, and so forth.
  - ❑ Availability of HN electrical power or alternate power sources.
  - ❑ Condition of road network, to include whether roads are dirt, cobble, improved, or paved.
  - ❑ HN maps of the exercise area.
- ❑ Critical construction issues for predeployment site-survey team. The team should obtain photographs of construction–related sites (if permitted by the HN), particularly base-camp areas, project sites, construction-material sources, haul roads, and so forth. Construction considerations that the survey team should address include-
  - ❑ Adverse construction conditions, such as swamps, rock, and extreme slopes.
  - ❑ Availability and location of construction water.
  - ❑ Distance from the base-camp site to construction sites.
  - ❑ Requirements to build construction roads, develop borrow pits or quarries, and so forth.

- ❑ Location of construction materials (raw materials and finished items, such as bridge components).
- ❑ Availability of HN–supplied materials, such as bridge components and well components (as agreed upon by the HN and the country team).
- ❑ HN capability to support procurement and delivery of typical construction materials expected to be used, to include estimated prices.
- ❑ Local availability, costs, and sources for plywood, 2– by 4–inch lumber, 4– by 4-inch lumber, and electrical materials that may be used in base-camp construction and facilities maintenance.
- ❑ Local availability, costs, and sources for construction equipment that may be leased.
- ❑ On-site survey data with elevations, dimensions, and stations.
- ❑ Soil analysis of all soils expected to be excavated or used in construction.
- ❑ Location of fences, structures, high–traffic roadways, or other obstacles that may affect construction.
- ❑ Formal route and bridge reconnaissance from the expected port of debarkation (POD) to the construction area.
- ❑ Specific locations and requirements for rights-of-way and easements to support the construction mission.
- ❑ Existing HN construction plans for the missions requested. Often these projects have been designed and on the books for years awaiting funding.
- ❑ Existing HN survey data for the construction site.
- ❑ HN hydrological studies, if available, for any rivers expected to be bridged or forded.
- ❑ HN cultural or ethic mores that would impact on construction; for example, occidental vice oriental latrines and geographic orientation of facilities.

❑ Potential for unconventional weapons in the area of operations. Determine if nuclear, biological, and chemical (NBC) protective masks and equipment will be used during the exercise. Determine if decontamination supplies and equipment are required.

❑ Request for early declassification of the exercise, to include deployment routes.

❑ Operations order (OPORD) format and/or requirements of the theater CINC.

❑ Supporting CINCs United States Army Forces Command (FORSCOM) or United States Southern Command (USSOUTHCOM) deployment order requirements.

❑ Restrictions, if any, on presence and/or participation of general/flag officers.

❑ Need for and source of translator/interpreters for the unit staff or duration staff (for ODT).

❑ Liaison party. Negotiate and/or request approval for the return of a small liaison team to the HN 3 to 6 days before the STARTEX. Include arrangements for transportation, quarters, and messing.

❑ Locations for separate headquarters, if desired, for the US service components. Determine collocation requirements for HN liaison officers at the US headquarters.

- ❑ Operational-equipment requirements for equipment that is not normally part of the unit, such as communications gear, medical equipment, and construction equipment.
- ❑ Predeployment training requirements for equipment not normally located in the unit, such as the tactical satellite (TACSAT), the Tactical Army Combat Service Support (CSS) Computer System (TACCS), and the Standard Army Retail Supply System (SARSS).
- ❑ Routes and modes of travel within the US. This includes the responsibility for transportation of nonorganic units.
- ❑ Designated ports and airports of departure and arrival. Coordinate with the transportation officer from the local installation, area support group, or Transportation Command (TRANSCOM), as appropriate.
- ❑ Restriction on weapons of heavy caliber (.50 caliber and above).
- ❑ Restriction on US flags or unit ensigns.

Intelligence-Considerations Checklist. The following considerations apply to intelligence planning:

- ❑ Current threat in the exercise area. If it is appropriate for the local environment, include threat engineer capabilities; for example, techniques they employ and types of mines they have. Determine resources to maintain up-to-date threat information; for example, human intelligence (HUMINT), image intelligence (IMINT), electronic intelligence (ELINT), medical intelligence, and local police.
- ❑ Contact with the US Defense Attache in the US embassy. Also contact the regional security officer, Chief of Station, and additional contacts with whom continuous liaison is required.
- ❑ Requirement for full or part–time counterintelligence liaison with the HN's intelligence services.
- ❑ Proposed location of tactical operations centers. If not collocated, determine communication interface requirements.
- ❑ Requirement for a sensitive compartmented information facility (SCIF).
- ❑ US government facilities that can receive, store, or send classified messages and documents. Determine if armed forces courier service and special security office are locally accessible in the exercise area. Determine if diplomatic pouch and back-channel service is available through the US embassy.
- ❑ Deployment sites classified storage and destruction requirements.
- ❑ Secure communications requirements; for example, fixed site, voice, teletype, and mobile.
- ❑ Areas that should be off limits to US forces. Notify the task force commander of these off–limit areas.
- ❑ Information on the attitude of the local populace toward US forces. Disseminate this within the task force.
- ❑ Opposing forces (OPFOR) uniform requirements, if applicable.
- ❑ Briefing room, wall board, projector, and security requirements for classified briefings, and so forth.

<u>Communication-Considerations Checklist.</u> The following considerations apply to the communications aspects of a deployment:

- ❑ Frequency requirements, frequency availability, and HN restrictions for—
  - ❑ Ultra high frequency (UHF).
  - ❑ High frequency (HF) and HF single side band (SSB).
  - ❑ Very high frequency (VHF).
  - ❑ Amplitude modulation (AM) and frequency modulation (FM).
  - ❑ TACSAT.
  - ❑ Teletype.
  - ❑ Secure communications equipment.
- ❑ HN communications equipment and interface requirements and restrictions.
- ❑ Communications requirements (secure and nonsecure) from entry points to the exercise site and from the exercise site to the US embassy, the theater CINC, CONUS, and so forth.
- ❑ Secure telephone unit (STU). Equipment is needed at both ends if secure telephone transmissions are necessary. Request keying materials through communications security (COMSEC) channels from the National Security Agency at least six months before departure.
- ❑ Requirements for receiving, storing, destroying, and accounting for COMSEC material, if applicable.
- ❑ Mobile communications requirements (secure and nonsecure) to support convoys, base camp, and work sites.
- ❑ Requirements for specialized communications equipment. Include any special support requirements, such as transformers, filters, and ground points.
- ❑ Communication-center security considerations, such as access controls, emergency destruction procedures, duress procedures, lighting, and special power requirements.
- ❑ Estimated cost of communications support, to include leased lines, commercial power, POL for generators, and transformers.
- ❑ Unique atmospheric phenomena that impact on communications.
- ❑ Communications systems that will be used by higher headquarters and other mandatory communications nodes. Ensure that the-task force ‘will have adequate assets to maintain continuous communications with these nodes.

<u>Personnel-Considerations Checklist.</u> The following considerations apply to the personnel aspects of a deployment:

- ❑ Rules on consumption of Class VI personal demand items (beer, cigarettes, and so forth) for exercise participants. Publish these rules within the task force.

- ❑ Establishment of standardized living conditions for all task force personnel, if possible. Inconsistent standards can cause morale problems within the task force.
- ❑ Impact of the US presence on the local populace in the deployment area. An influx of dollars (or HN currency) may have major repercussions in a poor area. Well–intended actions by US soldiers may have a negative impact on the local people's view toward the HN government.
- ❑ In-country records requirements, such as medical, dental, pay, and personnel records; passports; and job books.
- ❑ Country-unique medical requirements. In coordination with the medical staff, identify requirements and ensure that appropriate preventive measures (such as shots) are accomplished before deployment.
- ❑ Personnel requiring specialized medication. For individuals needing prescription medicines, ensure that they bring enough for the duration of their participation in the exercise, including any anticipated extensions and delays. Ensure that unusual medicine requirements are coordinated with the medical staff.
- ❑ Personnel requiring glasses. For those requiring prescription glasses, ensure that two serviceable pairs are available before deployment. If NBC gear is carried, ensure that a pair of prescription mask inserts are properly fitted to the individual's mask. It may be necessary to prohibit the use of contact lenses to avoid eye infections; coordinate this issue with medical personnel.
- ❑ Restrictions, if any, on female service members. For example, local customs in some Middle East countries can affect the duties of female personnel. In some cases, women are prohibited from driving and are required to wear long–sleeve garments. In some countries, officials may refuse to deal with females in leadership roles.
- ❑ Uniform and dress requirements, taking into account environmental factors.
  - ❑ Military on base.
  - ❑ Military off base.
  - ❑ Civilian dress requirements and restrictions.
  - ❑ Military en route to and from the exercise.
- ❑ Passport and visa requirements.
  - ❑ Try to obtain a waiver of passport and visa requirements. If approved, obtain approval or agreement for the use of a military identification (ID) card or a certified unit roster to satisfy HN requirements.
  - ❑ Try to obtain an entry fee waiver for deploying troops (for countries charging entry fees).
  - ❑ Determine if a need for HN ID cards exists.
  - ❑ Determine procedures required by the HN for in–processing.
  - ❑ Determine requirements for in-country travel.
  - ❑ Determine requirements for travel to major cities near the operational area (if different from above).
- ❑ Customs procedures for arrival and departure. Disseminate information on items prohibited by the US or the HN.

- ❑ In– and out-processing requirements once in country for AC and RC personnel.
- ❑ Country-clearance requirements for personnel entering the HN. If clearances are required, ensure that the request is processed in a timely manner. Establish procedures for obtaining country clearances for rotational personnel, visitors, and so forth.
- ❑ Procedures to be followed in the event of an accidental death.
- ❑ Charts of HN military ranks (officer and NCO).
- ❑ HN money system. Determine the US and HN money exchange rate and approved procedures.
- ❑ Dates of local holidays and routine hours of operation for local businesses/government offices. Determine the impact of these issues on task force operations. This may include the prohibition of equipment or haul operations, the closure of public and private offices that supply resources, and the absence of HN-contracted labor. Coordinate these issues with operations and logistics sections.
- ❑ Postal-service requirements. Determine whether the service will be for letters and packages or letters only. Determine procedures for the replenishment of stamps and other related activities, such as money orders.
- ❑ Support should be requested by a postal detachment, if warranted. Ensure that the detachment establishes an Army Post Office (APO) number before unit deployment.
- ❑ Emergency–leave procedures.
  - ❑ Determine the availability of Red Cross facilities in country.
  - ❑ Determine passport or visa requirements if travel is via commercial airlines.
  - ❑ Determine procedures for turn–in and reissue of HN ID cards if they are issued.
  - ❑ Establish procedures for tracking personnel movements between home station and the deployment site. A system must be established to maintain personnel accountability.
- ❑ Morale, welfare, and recreation support.
  - ❑ Determine recreational–facilities availability and access.
  - ❑ Determine and disseminate information on the use of cameras.
  - ❑ Determine if audio-visual equipment and movies can be obtained through the US embassy. If so, determine equipment requirements; for example, 8–mm versus 16–mm format.
  - ❑ Determine if personnel will be able to make phone calls to their home/home station.
- ❑ Financial–planning program. All soldiers and their families must be financially prepared for extended deployments. Checking or savings accounts must be established and allotments started so that families have access to funds at the home station. Deployed soldiers must take an adequate number of personal checks to cash during their deployment. Commanders must ensure that soldiers have planned for the welfare of family members.
- ❑ Basic load of tax forms and publications if the unit will be deployed during tax–filing time. Establish a training program for unit tax officers.

- ❑ Family–support program to assist the soldiers' families, These programs sustain morale and unit combat effectiveness. Key components of such a program include-
  - ❑ Information briefings and newsletters,
  - ❑ Guidance on powers of attorney and wills.
  - ❑ Unit and family assistance interface, including emergency–notification procedures.
  - ❑ Phone numbers for key POCs.

  (For more details, see the family support group checklist on page B-37.)
- ❑ HN Briefings for all personnel. All members of the task force should understand the culture, customs, traditions, and religion of the local people. They should also know the military and political goals of the exercise. An understanding of key phrases in the local language is also worthwhile.
- ❑ In-country briefing by the State Department or the HN for all deployed personnel. The liaison team should coordinate for this briefing.

<u>Public-Affairs-Considerations Checklist.</u> This checklist is based on the assumption that final, approved public affairs guidance for an exercise authorizes direct coverage of the exercise by news-media representatives. Also, there will be significant news-media interest by the American and foreign (including indigenous) press. Even if it appears clear that no on-scene media visits will be permitted, the information sought by this checklist should be obtained in case that policy changes on short notice. The following considerations apply:

- ❑ Principal and alternate public-affairs POCs within the US embassy.
- ❑ Principal and alternate POCs within the HN's Ministry of Defense and Ministry of Information for public–affairs aspects of the exercise.
- ❑ Number of news–media organizations represented in the country; for example, American, European, and indigenous.
- ❑ A list of media contacts in country, if available.
- ❑ HN's policy towards the foreign press corps.
- ❑ Requirement by the HN for accreditation of newsmen from outside the country.
  - ❑ Determine what documents are required, to whom correspondents must apply, and the approximate time to obtain accreditation.
  - ❑ Determine if the HN would object to news–media representatives accompanying exercise forces from CONUS aboard military aircraft.
- ❑ Standing Department of Defense (DOD), Joint Chiefs of Staff (JCS), and/or theater CINC press guidance and public–affairs policy for the exercise. Determine if it is compatible with the US Ambassador's public–affairs policy for the exercise and if any of the guidance is likely to change.
- ❑ HN's public-affairs policy for the exercise and whether it is likely to change.
- ❑ HN's ground rules for American and foreign news media. Determine if these rules are the same as those for indigenous media.

- ❑ Press release.
  - ❑ Determine if a press release has been drafted for the exercise.
  - ❑ Determine if the press release has been coordinated with the US embassy.
  - ❑ Determine if the press release and sample questions and answers have been coordinated with the HN.
  - ❑ Determine when the US embassy and the HN would like to see the press release published.
- ❑ Extent of US embassy public–affairs-officer (PAO) involvement in the exercise. Determine the US embassy ground rules for the task force PAO when dealing with news–media representatives and inquiries.
- ❑ CA personnel attached to the unit. If there are no CA personnel attached, the unit must identify personnel within the organization who speak the HN's language.
- ❑ Logistical requirements, such as transportation, food, and overnight accommodations for media visits to the exercise site.
- ❑ Policy and procedures for taking photographs, motion pictures, or videotapes of the exercise.
  - ❑ Determine HN sensitivities.
  - ❑ Determine if photographs, motion pictures, or videotapes may be released directly by the exercise PAO (after coordination with US embassy PAO) without clearance by the HN. If not, determine what arrangements must be made with the HN.
  - ❑ Give operations security (OPSEC) briefings to news–media personnel outlining reasons for security precautions and examples of Do's and Don'ts.
  - ❑ Determine requirements for local processing of film and distribution of prints.
- ❑ Facilities near the exercise site available for use as a press center in the event it is necessary to establish one.
- ❑ Requirements for access to the automatic voice network (AUTOVON) from the press center facility.
- ❑ Interpreter requirements for dealing with the indigenous press.
- ❑ Information and brochures, if available, on the closest suitable hotels that western news representatives could use.
- ❑ Special arrangements necessary for hometown news releases.

<u>Provost–Marshal–Considerations Checklist.</u> The following considerations apply to the security and law-enforcement aspects of a deployment:

- ❑ Effect that existing political agreements or Status of Forces Agreements (SOFAS) will have on enforcement activities. If a SOFA does not exist, determine what procedures will be followed.
- ❑ Appropriate weapons–, ammunition–, and equipment–security measures.

- ❑ Appropriate procedures for combatting terrorism. Determine sources of threat information. Establish special equipment requirements for sensors, alarms, communications, night–vision devices, and emergency lighting.
- ❑ List of prohibited items for customs processing. Disseminate the list and coordinate customs procedures for country arrivals and departures. Coordinate with customs inspectors for unit deployment procedures.
- ❑ A security concept that fits the situation.
  - ❑ Conduct overall physical security risk assessment for the command.
  - ❑ Determine the number of entry control points and area patrols at the base camp, deployed sites, airfields, and seaports.
  - ❑ Determine procedures for processing nonexercise participants for entry into exercise installations and sites, including official and unofficial visitors.
  - ❑ Determine requirements for combined-security arrangements.
  - ❑ Determine the jurisdiction of HN police and the type of jurisdiction of task force security personnel.
  - ❑ Determine interface requirements, to include joint patrols and bilingual guards.
  - ❑ Determine enforcement requirements due to local customs and values, for example, photography, consumption of Class VI items, and dress requirements.
  - ❑ Determine off–limits establishments and distribute information.
  - ❑ Determine procedures to control local vendors at US encampment sites.
  - ❑ Determine augmentation requirements and procedures when military police units rotate.
  - ❑ Determine security procedures when other major task force personnel rotations occur.
- ❑ Measures that ensure the security of government and private property. Address measures to be taken during all phases of the operation; for example, during transit, in ports of embarkation and debarkation, in the base camp, on project sites, and during convoys.
- ❑ Briefing for deploying troops. Include SOFA cards that contain instructions and POCs for troops experiencing problems with local authorities. The card should contain the information in English and the language of the HN.

<u>Medical–Services–Considerations Checklist.</u> The following considerations apply to the medical–services aspects of a deployment:

- ❑ General medical information for area of operations.
  - ❑ Endemic and epidemic diseases prevalent in the area of operations.
  - ❑ Poisonous plants, wild animals, and reptiles (land and water).
  - ❑ Domestic animals and animal diseases.
  - ❑ Pest management.

- ❑ Health status of the general populace.
- ❑ Availability of local medical and dental care for the general populace.
- ❑ Availability of local veterinary programs.

❑ Health service support considerations for deploying forces.

- ❑ Immunizations and chemical–prophylaxis requirements.
- ❑ Location and capability of supporting medical element (coordination requirements if medical support is being provided by another US service, allied nation, or HN, including emergency medical treatment and hospitalization).
- ❑ Availability of medical evacuation, to include road networks, aeromedical–evacuation resources and procedures, and communications requirements.
- ❑ Preventive-medicine programs and measures requirements, to include inspection and approval of potable-water sources, dining–facility inspections, field hygiene and sanitation, and pest management.
- ❑ Requirements for field-sanitation teams, to include training and equipment.
- ❑ Veterinary assets for the inspection of Class I items for wholesomeness and for approval of local sources for foodstuffs.
- ❑ Personal–protective supplies, such as insect repellant, mosquito bars, sunscreen, tropical or cold weather clothing, and protective masks for dust.
- ❑ First-aid refresher training and combat-lifesaver training for deploying forces.
- ❑ Training and equipment requirements.
- ❑ Level of basic supplies that each deploying unit should bring; for example, 5 days, 15 days, 30 days, 60 days, or more, based on table(s) of organization and equipment (TOE) requirements and on operational contingency. Determine Class VIII resupply requirements and procedures and the location of the supporting facility.
- ❑ Medical-communications support requirements for the base camp to CONUS and dispersed sites, medical readiness teams, and so forth.
- ❑ Development of a contingency plan for mass-casualty situations.
- ❑ Coordination requirements for civic action programs that include medical operations (refer to Field Manual (FM) 8–42 for medical–mission assessment information).
- ❑ Distance from the base camp to where medical civic action program activities are to be conducted.
- ❑ Accessibility of the area of operation, such as paved roads, paths, pack animals, ground transportation, or air transportation requirements.
- ❑ Types of medical missions based on medical needs of the local populace.
- ❑ Climatic– and environmental–related threats to health.
- ❑ Operation length.
- ❑ Participation level of HN (civilian or military) health professionals.
- ❑ Local sources of medical materials.

Religious-Program-Considerations Checklist. The following considerations apply to the religious program of a deployed unit:

- ❑ Worship environment.
  - ❑ Identify the principal faith groups represented in the HN.
  - ❑ Identify any restrictions on, or prejudices relating to, the free exercise of religion.
  - ❑ Identify the name and location of the senior HN military chaplain holding jurisdiction in the exercise area.
  - ❑ Identify the name and location of the senior US military chaplain in the HN, if any.
  - ❑ Identify the name and location of any US and foreign missionaries near the exercise area.
  - ❑ Identify any worship facilities, military or civilian, within or near the exercise area.
  - ❑ Determine if a worship facility, other than a general messing facility used outside of meal times, is necessary for exercise participants.
  - ❑ Ensure continuous chaplain coverage for the exercise.
- ❑ Counseling support.
  - ❑ Identify the nearest overseas telephone terminal.
  - ❑ Identify the availability and the location of a Military Affiliate Radio System. (MARS) station.
  - ❑ Identify the nearest POC for the Red Cross.
  - ❑ Identify procedures for emergency–leave travel.
- ❑ Possibilities for civic action.
  - ❑ Identify orphanages, schools, child-care centers, nursing homes, and so forth near the exercise area that might benefit from volunteer civic action.
  - ❑ Identify the POC for civilian social services in the exercise area, if any.
  - ❑ Provide collection points for food, clothing, and other donated items. Identify recognized, local organizations that may distribute these items.

Legal-Considerations Checklist. The following considerations apply to the legal-services aspects of a deployment:

- ❑ Exercise-training limits, such as permissible and forbidden activities; geographical limits for CA, medical, and engineer personnel. Determine these limits through the appropriate Judge Advocate General (JAG).
- ❑ HN support agreements, protocols, memoranda, and similar documents. Obtain these documents from the US embassy in the HN.

- ❑ SOFA. Determine if a SOFA is in effect. If not, coordinate requirements with the JAG, the theater CINC's Staff, and country team personnel in the US embassy. Potential items include-
  - ❑ Privileges and immunities of US soldiers during the exercise in the HN.
  - ❑ Crimes committed by a US soldier. Determine whether the HN has jurisdiction to try the individual.
  - ❑ Concurrent jurisdictions. If the US and the HN both have jurisdiction to try criminal offenses, determine which nation will try the cases and what procedures will be followed.
  - ❑ Procedures for obtaining custody of US personnel incarcerated by the HN police and what reports are required.
  - ❑ Tax or customs liabilities of US personnel entering or leaving the HN.
  - ❑ Whether US personnel are subject to civil suits. If so, obtain a compendium of civil actions that might arise.
- ❑ Procedures, rules, or regulations that must be followed by US personnel bringing personal property into the HN.
- ❑ Restrictions, if any, on motor vehicle operations by US personnel.
- ❑ Financial obligations incurred by the US through exercise participation in the HN.
- ❑ Duties and obligations the US has in regard to military equipment brought into the HN.
- ❑ Duties and obligations that will be imposed on the US for facilities that they construct.
- ❑ Assumed or agreed to obligations of the HN in regard to all phases of support.
- ❑ HN policies on whether US authorities can employ local nationals.
- ❑ Infringements on HN sovereignty.
- ❑ Claims.
  - ❑ Determine procedures for in-country claims if filed against the US government and what documentation is required from those filing the claim.
  - ❑ Determine what claims procedures the US has agreed to for claims filed against the HN.
  - ❑ Determine if a claims specialist will be assigned to the task force. If not, determine who will handle claims.
  - ❑ Provide exercise participants with information on key Do's and Don'ts so that claims against the US government and/or the individual can be avoided.
- ❑ Authority of the task force chain of command under the Uniform Code of Military Justice (UCMJ). Determine who has court–martial convening authority for the various levels of courts–martial.
- ❑ Court–martial arrangements to bring in a judge for courts–martial or to transport soldiers and witnesses to court.

- ❑ Requirements for Trial Defense Service (TDS) support. Establish procedures for providing support to the task force, such as rotational deployment of TDS personnel, transportation of unit personnel to see TDS lawyers, or other alternatives.

Logistical-Considerations (Facilities) Checklist. The following logistical considerations apply to facilities planning for a deployment:

- ❑ Billeting.
  - ❑ Determine billet availability and procedures for requesting them through HN military and commercial sources.
  - ❑ Determine how the troops will be billeted in the base camp; for example, tents, temporary huts, or portable shelters.
  - ❑ Assess cost and relative safety of electric, coal, wood, oil, and other types of heaters if heating is required for billets.
  - ❑ Determine billeting spaces required for overnight and VIP guests.
- ❑ Work structures.
  - ❑ Determine the requirement for and the availability of environmentally controlled structures for facilities such as the tactical operations center (TOC), aircraft and equipment maintenance shelters, and communications centers.
  - ❑ Determine the requirement for and the availability of other work areas through HN military and commercial sources.
  - ❑ Determine request procedures for HN military and commercial sources.
  - ❑ Determine the availability of motor pools with security fences and lighting.
- ❑ Ammunition storage facilities.
  - ❑ Locate an ammunition storage point if live ammunition is deployed.
  - ❑ Determine storage security requirements.

Logistical-Considerations (Supply) Checklist. The following logistical considerations apply to supply-support planning for a deployment:

- ❑ Supply-source standard operating procedures (SOPs), documentation, and accounts. Ensure that the advance party has an adequate number of supply personnel to open accounts, to request and receive supplies, and to initiate contracts.
- ❑ Procedures for maintaining property accountability. These procedures are especially important as property is transported from home station to the exercise site and when the property is being redeployed. Equipment and sets, kits, and outfits (SKOs) deploying early for base-camp construction and returning late after site breakdown must be closely controlled.
- ❑ Class II, III (package), VI (sundry packs-male and female), and IX. Establish operational packages to support operations until the in-country logistical support base is established.

- ❑ Procedures for the replacement of components of SKOs. Shortages of engineer hand tools and mechanics tools can have a serious impact on unit operations.
- ❑ Requirements for initial self-service supply centers (SSSC) stockage. Establish accounts for SSSC replenishment in country. In some cases, it maybe necessary to replenish SSSC items from the home station.
- ❑ Requirements for environment-specific items.
  - ❑ Suntan lotion (sunscreen).
  - ❑ Sunglasses and goggles.
  - ❑ Insect repellent.
  - ❑ Mosquito net.
  - ❑ Water–purification tablets.
  - ❑ Hot– or cold–weather uniforms and accessories.
- ❑ POL procedures.
- ❑ Determine the availability, source, and distribution of POL; for example, JP–4, JP-5, JP–8, motor gasoline (MOGAS), DF2, alternative fuels, and POL additives for that climatic region.
  - ❑ Determine support requirements for petroleum storage and distribution.
  - ❑ Ensure that HN equipment is interchangeable with US equipment.
  - ❑ Provide for POL quality control. Task for a lab, if necessary. If fuel is provided by contract, ensure that the contract specifies appropriate US quality-control standards.
  - ❑ Ensure that the dispensing unit can filter the fuel and that the task force deploys with extra fuel filters if only poor–quality fuel is available.
  - ❑ Determine HN transportation and distribution capabilities.
- ❑ Requirements for Class IV barrier material for the base camp, motor parks, job sites, and so forth. Determine sources of supply for Class IV items. In some areas, a rotating stock of Class IV items is available to deploying units.
- ❑ Procedures to control guard ammunition. Establish procedures for issuing basic load of ammunition, if required.
- ❑ Procedures for transferring equipment. During some exercises involving two-week rotations of units, an equipment pool will be used. Establish a task force equipment identification system.
- ❑ Procedures for maintaining property accountability during marshaling operations.
- ❑ Army and Air Forces Exchange Service (AAFES) field concession. If a field concession is needed, establish appropriate stockage items and resupply procedures.
- ❑ Procedures for initiating and processing reports of survey.
- ❑ Requirements for fire-fighting equipment and HN support capabilities.
- ❑ Requirements for a logistical liaison team from the theater CINC.

- ❑ Availability of commissary privileges in the US embassy. This is important for the task force commander who entertains VIPs (US and HN) at the task force site.
- ❑ Coordination between rotations. This is essential during exercises in which the logistical support element (LSE) for the task force includes rotational personnel. Duration personnel should be assigned to key LSE positions, if possible. A continuity file should be developed to address problems, lessons learned, and recommendations for corrective actions.
- ❑ Libraries of appropriate doctrinal publications and policies and procedures as well as any contingency plans that may apply to the operation.
- ❑ Plans for phased deployment of logistics assets. Ensure early deployment of assets to provide mess capability, water production, fuel supply, Class IV control, and exercise property distribution.

Logistical–Considerations (Services) Checklist. The following logistical considerations apply to services-support planning for a deployment:

- ❑ Messing.
  - ❑ Determine if messing can be arranged in the local area (clear with medical staff). If not, determine total messing facility requirements.
  - ❑ Determine the ration cycle that is supportable by ration–issue personnel and available equipment.
  - ❑ Determine if rations can be supplemented with food purchased locally.
  - ❑ Determine refrigeration and other special processing requirements. Refrigeration requirements must be thoroughly planned to ensure sufficiency. Consider the size of the force supported, the type and reliability of refrigeration equipment, the ration cycle, methods of ration transport, and the distance rations must be transported.
  - ❑ Coordinate with veterinary personnel for the inspection of Class I items for wholesomeness,
  - ❑ Coordinate with preventive-medicine personnel for sanitary inspections of dining facilities.
- ❑ Water.
  - ❑ Determine requirements for water-storage support.
  - ❑ Determine the availability and sources of potable water.
  - ❑ Determine the quality and quantity of potable water needed as well as storage requirements and methods.
  - ❑ Determine requirements for water-purification support.
  - ❑ Determine HN transportation and distribution capabilities.
  - ❑ Determine the availability and the proximity of wells and the equipment required to use them.
  - ❑ Determine the need for water containers, bladders, and/or coolers.
  - ❑ Determine the availability and sources of ice.

- ❑ Determine the quality and quantity of ice required.
- ❑ Determine required health test equipment and procedures.
- ❑ Coordinate with preventive-medicine personnel on the inspection of water sources, water, and ice for potability.

❑ Hygiene.

- ❑ Determine the availability of laundry services in the HN.
- ❑ Determine the availability of shower facilities. Request bath and laundry units, as required.
- ❑ Determine if portable latrine facilities can be contracted in the HN. If not, determine latrine and waste disposal procedures and the capacity needed.
- ❑ Determine the method for trash and garbage disposal.

❑ Electric power.

- ❑ Determine the projected base-camp power demand.
- ❑ Determine alternative power sources if the HN cannot meet requirements.
- ❑ Determine the availability of HN commercial electrical power.
- ❑ Determine any transformation and switchgear needs if HN power is available.
- ❑ Determine the reliability of HN power and whether any backup generation capability is required.

Logistical-Considerations (Transportation) Checklist. The following logistical considerations apply to transportation planning—

❑ For the predeployment phase of a deployment:

- ❑ Applicable transportation references. Ensure that these references are available for planning and conducting unit movements.
- ❑ Identification and certification of hazardous materials.
- ❑ Hazardous cargo required for various types of deployments should be identified and certification information made readily available. All necessary packing materials to meet performance oriented packaging (POP) standards should be readily available.
- ❑ Air–movement training. Ensure that the unit has an adequate number of air–load planners and hazardous cargo certifiers. FORSCOM Regulation 55–1 requires that each unit and intermediate command have at least one unit–movement officer appointed on orders. At least one person in each unit should be qualified to certiify hazardous-cargo air shipments.

❑ For the deployment and redeployment phases of a deployment:

- ❑ Basic considerations. The general concept for movement must include the following information and features so that comprehensive planning can be conducted. This information must be coordinated with the logistical support unit early in the planning process—

- ❑ Determine how equipment will be moved to the sea port of embarkation (SPOE).
- ❑ Determine how personnel will be moved to the aerial port of embarkation (APOE).
- ❑ Determine what equipment should be on the initial sea–lift list.
- ❑ Determine the number of personnel to be moved by airlift.
- ❑ Determine what equipment is to be airlifted.
- ❑ Determine target dates for each movement.
  - ❑ Reconnaissance party.
  - ❑ Advance-party equipment.
  - ❑ Advance-party personnel.
  - ❑ Main–body equipment.
  - ❑ Main–body personnel.
- ❑ Determine SPOE. Determine the capacity for ships, drafts, dock space, and lift equipment.
- ❑ Determine APOE.
- ❑ Determine sea ports of debarkation (SPODs). Determine the capacity of each SPOD for ships, drafts, dock space, and lift equipment.
- ❑ Determine aerial ports of debarkation (APODs).
- ❑ Determine trail party personnel and equipment for redeployment only.

❑ Convoy routes from points of debarkation, arrival ports, and airfields to base camp or field training exercise (FTX) area.

- ❑ Submit convoy requests.
- ❑ Mark vehicles.
- ❑ Check placards.
- ❑ Prepare strip maps (include checkpoints).
- ❑ Verify convoy clearances.
- ❑ Establish and check communications.
- ❑ Coordinate lodging.
- ❑ Coordinate messing.

❑ Coordinate safety plans with civil authorities, such as police, for traffic control points.

- ❑ Coordinate with PAO.
- ❑ Set speed limitations.
- ❑ Plan rest stops.

❑ Establish refueling locations (address environmental restrictions and considerations).

- ❑ Coordinate remain overnight (RON) locations.
- ❑ Check road conditions.
- ❑ Assess security considerations.
- ❑ Develop controls for critical and pilferable materials.

- ❑ Supercargo to accompany sea–lift cargo. Select these personnel based on military occupational specialty (MOS)–related tasks that will be performed aboard the ship.
- ❑ Vehicle packing. Vehicles may be loaded with ammunition basic load, POL, or other equipment. If so, ensure automated unit-equipment-list (AUEL) data and other shipping data reflect accurate weights.
- ❑ Bar code labels. Labels should be affixed to all vehicles and equipment for use with automated transportation systems.
- ❑ Customs and shipment of vehicles and equipment. Ensure that fueling and defueling is planned for by the departure/arrival airfield control group (D/AACG). Determine US Department of Agriculture shipping requirements, and if special hoses, attachments, or steam cleaners are required to clean vehicles. Identify wash facilities at or near SPOD and/or APOD for customs inspection. Schedule wash racks or commercial facilities as needed.
- ❑ Materials-handling-equipment (MHE) requirements and availability. Determine if International Standardization Organization (IS0) containers can be used from port to base camp.
- ❑ Wrecker–support requirements and availability.
- ❑ Passenger– and cargo-processing and holding–area requirements.
- ❑ Tasking level for transportation maintenance and repair requirements (unit, intermediate, or depot).
- ❑ Liaison with each port through which personnel and equipment will be deployed. Coordinate with D/AACG. Schedule predeployment conferences with port personnel during the preparation phase of the exercise. Determine any local requirements for the use of only HN personnel and whether they are qualified to operate Army heavy equipment.
- ❑ Unit transportation liaison. Identify and establish liaison with the movement control team/organization that will support your unit/mission in theater. Provide unit liaison teams to all aerial and sea ports of embarkation and debarkation with detailed deployment information, including the unit's deployment equipment list (DEL), movement information, hazardous-cargo data, and vehicle-key-control procedures. The port liaison team must have authority to make decisions based on established milestones.
- ❑ Unit load teams. Establish unit load teams to stage and prepare loads for shipping. Unit load team duties include–

- ❑ Transporting equipment to sites (unit staging area and then to transportation node).
- ❑ Performing organizational maintenance at the unit staging area or transportation node, if necessary.
- ❑ Configuring and labeling loads for shipment at the unit staging area.

- ❑ Loading and securing vehicles.

- ❑ Rail loading. Ensure that ramps are ready for unloading and off-loading equipment. Ensure that rail-load teams have adequate blocking, bracing, and tie-down material. Vehicles without hardened-steel, tie-down shackles (clevis and pin) are nondeployable. Have shackles installed before load out. Shackles are highly pilferable, so carry extras.
- ❑ POL for equipment transported by ship. Ensure that a POL point is established at ports where equipment will be off-loaded. Shipping restrictions limit the amount of fuel that maybe left in fuel tanks. Vehicles must be topped off once they are unloaded.
- ❑ Support for in–transit personnel. Ensure that liaison teams, load teams, and so forth are provided with rations, water, latrines, and shelter (and billeting if appropriate) at their work sites. Establish contact with the area support group (ASG) and local agencies to coordinate support for soldier health and welfare.

❑ For the execution phase of a deployment:

- ❑ HN transportation, Determine in-country transportation systems that will support mission requirements. Determine the systems' capabilities (during dry or wet weather) and restrictions (maximum size or gross weight of vehicles).
  - ❑ Seaports.
  - ❑ Inland waterways.
  - ❑ Roads.
  - ❑ Railroads.
  - ❑ Logistics over the shore (LOTS).
  - ❑ Airfields.
  - ❑ DZs.
- ❑ Durability of main supply routes. Determine the number of routes and who will be using them. Determine if the task force will need to dedicate effort to maintaining local main supply routes. In many regions, the number of paved roads suitable for heavy traffic is very limited.
- ❑ Vehicle-operator liability. Determine if US forces are protected from civil suits in case of a vehicle accident, either on or off duty. If not, determine the availability and cost of accident insurance.
- ❑ Troop transportation. Determine if trucks and/or buses are available locally for troop movements and who provides the drivers.
- ❑ Special requirements in area of operations. Identify any special items required to operate vehicles; for example, special tires for use in sand. Adjust authorized stockage list (ASL) and prescribed load list (PLL) quantities accordingly. Identify special equipment the unit must provide, such as aircraft tie-downs.
- ❑ Requirements for international drivers licenses or other special driver training. Determine if there are any unique local driving laws.

- ❑ Resupply transportation requirements. Based on anticipated quantities of resupply items, determine requirements for vehicles to transport supplies from major drop-off points to task force sites. Determine the amount of items to be moved forward by helicopter versus truck or pack animals.
- ❑ Wrecker–support requirements and availability.
- ❑ Tasking level for various transportation maintenance and repair requirements (unit, intermediate, or depot).

Logistical-Considerations (Contracting) Checklist. The following logisitical considerations apply to planning for contracting during a deployment:

- ❑ Sufficient field-experienced contracting agents should be attached to and collocated with the task force. These officers should have prepared contingency contracting–support kits which include data on local resource availability, as well as potential host–nation-support (HNS) assets. Ensure that they will be able to write contracts covering every phase of the exercise. If possible, they should understand the language and customs of the HN. Determine if it is necessary for contracting officers to wear civilian clothing.
- ❑ Requirements for contracts. Determine what items and services should be contracted. Determine bilingual (oral and written) support requirements. Determine how funding will occur and, where applicable, estimate the cost of the following contract items:
  - ❑ POL, price per unit of measure (liter or gallon).
  - ❑ Water, price per unit of measure.
  - ❑ Electricity, price per kilowatt–hour (commercial power backup and use).
  - ❑ Officer, enlisted, and female quarters.
  - ❑ Engineer services, such as the disposal of garbage and sewage and the preparation of land areas, for example, leveling, ditching, sumps, and so forth.
  - ❑ Messing and dining facilities.
  - ❑ Facilities repair and maintenance.
  - ❑ Land and facility rental or usage charges.
  - ❑ Equipment rental.
  - ❑ Vehicle rental.
  - ❑ Mobile-electric-power-source rental.
  - ❑ HN civilian hire.
  - ❑ Repair parts procured locally.
- ❑ Imprest funds. Determine how much money in imprest funds is required to support the exercise.
- ❑ Items that may be considered for contracting include-
  - ❑ Construction-equipment leases for unique (non–modification table of organization and equipment (MTOE)) equipment.
  - ❑ Base-camp materials.

- ❑ Rental vehicles, such as cars, four–wheel-drive vehicles, commercial vans, and refrigerated vans.
- ❑ Laundry services.
- ❑ Fresh fruits and vegetables. The source for these items must be approved by a veterinarian.
- ❑ Lodging for personnel providing liaison to the US embassy.
- ❑ Heavy haul of equipment (rail or truck).
- ❑ Trash removal and waste burning.
- ❑ Fuel coupons for rental vehicles.
- ❑ Fuel tankers.
- ❑ Bulk POL.
- ❑ Bus transportation.
- ❑ Potable ice.
- ❑ Oxygen and acetylene.
- ❑ Fumigation of the base camp.
- ❑ Steam cleaners.
- ❑ Newspapers.
- ❑ Repair parts.
- ❑ Forklift support.
- ❑ Tire changing.
- ❑ Machine-shop service.
- ❑ Telephone service.

❑ Maintenance of contracted equipment. Ensure that contracted equipment comes with service agreements. Establish a clause for an equipment contractor to repair equipment at the site where it is being used by US forces. Establish in-country service or parts contracts for equipment contracted in CONUS that does not include service agreements.

❑ Accountability of contracted equipment. Establish procedures for maintaining accountability of contracted equipment; for example, rental cars and construction equipment. Establish procedures for tracking scheduled maintenance on contracted equipment (contractor– or unit–performed).

❑ Coordinate closely with local authorities to locate the boundaries of property leased for base-camp sites, quarries, project easements, and so forth. This coordination is necessary to prevent encroachment on land that has not been leased. Boundaries should be plotted on survey maps of sufficient scale to prevent infringement of private-property rights. Determine restrictions on construction, digging, discarding refuse, and erecting towers.

❑ Existing facilities that maybe leased. Determine if an existing facility meets task force needs when supporting a unit requirement. For example, leasing an abandoned mining camp versus leasing an area and building a base camp.

Logistical-Considerations (Maintenance and Repair Parts) Checklist. The following logistical considerations apply to planning for maintenance and repair parts operations for a deployment:

- ❑ Predeployment equipment inspections. Ensure that all equipment being deployed for an exercise receives a technical inspection (TI) by qualified personnel before deployment. This inspection is necessary to avoid shipment of nonoperational equipment. Particular attention should be given to equipment not normally used at home station, such as bakery units.
- ❑ Advance party. Ensure that the advance party has an adequate number of mechanics, tools, and licensed operators to receive and move unit vehicles at the port.
- ❑ Procedures for local purchase of repair parts. Determine procurement procedures, identify local sources of repair parts, and determine requirements for training Class A/contracting personnel.
- ❑ Maintenance-support requirements and responsibilities. Identify all CONUS–based, in-country US, and local contract sources of support. Ensure that support is provided for all phases of the operation, including deployment and redeployment.
- ❑ External maintenance support. Determine requirements for maintenance support beyond task force capabilities. Establish support agreements and open accounts with other units that can provide the support, including in-country, in–theater, or CONUS–based units.
- ❑ Contact maintenance. Establish procedures for contact maintenance. Contact–maintenance teams can quickly repair minor deficiencies. Lube and service units positioned at key locations can decrease turn around time for vehicle maintenance.
- ❑ Low-density equipment. Determine which parts have a low density in the supply system so alternative sources and finding methods can be established.
- ❑ Commercial equipment. Determine procedures for using the commercial-operated parts depot system (COPADS) and commercial-operated parts retail systems (COPARS). Determine the availability of repair parts through local sources, such as John Deere, Caterpillar, and Case dealers.
- ❑ Computer maintenance. Determine computer (hardware and software) and other specialized maintenance requirements.
- ❑ Maintenance of temporary–loan equipment. Determine maintenance requirements for deployment–unique equipment, such as power generation equipment and refrigerator vans. Equipment failure of these critical low-density items can have a serious impact on the deployment.
- ❑ Aviation maintenance. Determine the tasking level for aviation–maintenance requirements and procedures (unit, intermediate, or depot).
- ❑ Warranty repairs. Determine how equipment repairs under warranty will be handled.
- ❑ Repair–parts requisition procedures (manual versus computer system).
- ❑ Force/activity designator (FAD). Determine if a temporary upgrade is required.
- ❑ Project codes. Determine if project codes for the exercise have been established.
- ❑ DOD activity address codes (DODAACs). Determine if in-country DODAACs are required.

- ❑ Order ship time (OST). Determine the estimated OST for the exercise.
- ❑ Exercise-specific repair parts. Determine repair–parts requirements and order them accordingly. Factors that impact on these requirements include the increased use of equipment, environmental considerations, length of lines of communication (LOC), and local availability of repair parts. Vehicle transmissions and tires are particularly critical in undeveloped areas.
- ❑ Back orders. Determine how back-ordered requisitions will be tracked. Determine cut-off dates and cancellation procedures for orders to close the supply pipeline before redeployment.
- ❑ Equipment transfers and repair parts. Establish procedures to ensure that repair parts, including major assemblies, for equipment transferred from one unit to another unit follow the equipment. These procedures are particularly important in exercises that have two-week unit rotations, as equipment will possibly be cross leveled based on unit size.
- ❑ Requirements for emergency resupply. Determine methods of delivery-airdrop, air, land, sea, and helicopter.
- ❑ Unserviceable major end items. Determine procedures for disposal of unserviceable equipment (HN versus stateside disposal). Determine procedures to establish a cannibalization point for wrecked vehicles.
- ❑ Facilities. Determine maintenance and repair–parts storage space requirements and any environmental constraints.
- ❑ Commercial power. Determine availability and compatibility of HN electrical power and whether any equipment has special electrical-power requirements.
- ❑ Dispatching. Establish project dispatching procedures or validate existing procedures.
- ❑ Licensing. Ensure that an adequate number of licensed personnel are designated to deploy to support the exercise.
- ❑ Publications. Determine technical manual (TM) requirements and procedures. Manuals may stay with vehicles for the duration of the deployment or each rotation may bring their own set of manuals.
- ❑ Tools. Determine what repair tools are required. Determine who will maintain them—the task force duration personnel or each rotation.
- ❑ Test, measurement, and diagnostic equipment (TMDE). Determine TMDE–calibration support requirements and who will perform them.
- ❑ Army Oil Analysis Program (AOAP). Determine the availability of laboratory facilities for oil analysis. Establish procedures for receiving timely test results at the deployment site.

Comptroller-Considerations Checklist. The following considerations apply to the budgeting aspects of a deployment:

- ❑ Budget implementation. Determine funding sources and authorizations for the operation. Determine who will monitor costs in the various categories, such as transportation, POL, resupply, and operations and maintenance.
- ❑ Items that are to be provided by the HN. Establish procedures for reimbursement to the US and requirements for maintaining accountability and records.

- ❑ Funding sources, accountability procedures, estimated costs, and so forth, used for the following types of items:
  - ❑ Rental vehicles and equipment.
  - ❑ Utilities.
  - ❑ Engineer services, such as garbage and sewage disposal, and base-camp-site preparation.
  - ❑ POL.
  - ❑ Soldiers' salary while in country.
- ❑ Contracting. Determine if an on-site agent has the authority to certify the availability of funds. If not, determine procedures to ensure the timely execution of task force contract requirements. Estimate tentative contracts, to include–
  - ❑ Dollar amount.
  - ❑ Purpose.
  - ❑ Contractor selection.
  - ❑ Bilingual-(oral and written) support requirements.
- ❑ On-site fund requirements.
  - ❑ Determine how local transactions will be accomplished-ash, credit card, check, or another method.
  - ❑ Estimate funds required by imprest–fund agents (cash and obligation authority).
  - ❑ Establish procedures and the source for funds conversion. If the US embassy is the source, determine what notice or information they will need for fund conversion.
  - ❑ Establish a pay and per-diem-processing policy and procedures for in-country payments.
  - ❑ Establish a policy for emergency funds, to include limits on how much cash may be stored. Determine requirements for funds storage locations and accountability controls.
  - ❑ Instruct Class A agents, purchasing officers, and ordering officers on the specifics and restrictions of each job.

<u>Aircraft-and-and-Airspace-Considerations Checklist.</u> The following considerations apply to the aviation aspects of a deployment:

- ❑ Coordination procedures with the Military Airlift Command (MAC) for troop-airlift support requirements. Establish lead times and cutoff points for changes. Determine a consolidated preparation for oversea movement (units) (POM) and a preparation of replacements for oversea movement (POR) site where airlift support is available.
- ❑ Air-cargo loading requirements and procedures. Determine whether air–terminal movement-control teams or unit personnel perform air-cargo loading.

- ❑ Coordination with MAC for overflight and refueling agreements. Coordination is required to ensure that agreements are made with appropriate HNs depending on the route of travel.
- ❑ Identification, friend or foe (IFF) considerations for local surface-to-air missiles or radar.
- ❑ Parking space for aircraft (coordinate with components and MAC) at the POM and POR sites, as well as potential HN sites.
- ❑ Air-traffic-control procedures within the HN, such as control-tower manning (US or HN only or combined responsibility). Determine flight-plan filing requirements, radar-traffic-control procedures, and so forth. Request an air–traffic-control (ATC) team for bare-base operations.
- ❑ Number, types, and locations of helicopters and pads required for the exercise. Determine refueling procedures and fire-fighting equipment needs. Determine special training requirements; for example, sling loads, high–altitude operations, and border–trace familiarization. Determine requirements for airfield matting or dust control agents to support aircraft operations.
- ❑ Twenty–four hour flying operations. Determine if they will be conducted, and if so, whether there are any special coordination requirements.
- ❑ Transportation for HN personnel. Determine if legal release forms are required for HN personnel to ride in US helicopters.
- ❑ Procedures for monitoring and scheduling movement of equipment and personnel via airlift at in-country locations. Determine communications procedures and requirements so that the task force can stay abreast of aircraft schedule changes.
- ❑ Ground-coordinate accuracy requirements and data source for aligning inertial navigation systems.
- ❑ DZ for airborne operations requirements. Determine zone safety requirements and HN coordination procedures.
- ❑ Air resupply.
  - ❑ Resupply may be conducted by airdropping or off loading supplies near the exercise site with subsequent helicopter or truck movement to the exercise supply sites. If intermediate off–loading is planned, determine MHE requirements.
  - ❑ Determine if precision controlled–approach radar or an instrument landing service is available if air resupply is critical. If not, determine if mobile systems can be obtained and certified to support the exercise.
- ❑ Refueling requirements at HN sites. Determine fire-fighting equipment and personnel requirements. Determine how US quality-control standards for POL will be met.
- ❑ Weather support. Determine who will provide weather information to the aircrews.
- ❑ Backup support. Establish alternative procedures for situations when adverse weather is anticipated and aircraft are unable to fly. These procedures should address ground resupply, troop transport, and medical evacuation (MEDEVAC).

- ❑ Command and control procedures for aircrews. Determine what procedures are needed for the task force to communicate with aircraft while en route from CONUS or intermediate points. Determine how the task force will stay in the information loop should an air diversion become necessary due to weather or aircraft system problems.
- ❑ Procedures for obtaining unscheduled air support in the event of a change to mission requirements.
- ❑ Procedures for aircraft security at normal operating locations (HN, US, or combined responsibility). Determine security and maintenance procedures in the event of unscheduled diversions.
- ❑ Billeting provisions for aircrew rest requirements.
- ❑ Maintenance and spare-part requirements for assets deployed in-country. Determine environmental constraints, the need for precision measuring equipment, special laboratory support requirements, and so forth.
- ❑ Transportation of aircraft to area of operations. Determine if aircraft will self–deploy. Use cocoon wrap to protect aircraft from salt water on sea deployments.

Base-Camp-Considerations Checklist. The following considerations apply to base-camp planning and construction:

- ❑ TOC area.
  - ❑ Perimeter concertina wire.
  - ❑ General purpose (GP) small tents for commander, S3, and VIPs.
  - ❑ GP medium tents for briefing areas, operations sections, communications sections, and weapons security.
  - ❑ Guard shacks at entrances. Provide access rosters to the security unit at secure area entrances.
  - ❑ Sandbag and gravel sidewalks.
  - ❑ Tent floors or covers.
- ❑ Troop living area.
  - ❑ Sleeping tents.
  - ❑ Orderly rooms.
  - ❑ Supply rooms.
  - ❑ Supporting units.
  - ❑ HN forces.
- ❑ Communications. An area for microwave tower or tactical-communications equipment should be secured (will be remoted to the TOC).
- ❑ Staff and administrative areas.
  - ❑ Personnel Officer's (S1's) tent.
  - ❑ Supply Officer's (S4's) tent.

- ❑ S4 yard with security fence and security lighting.
- ❑ Weapons storage areas; for example, container express containers (CONEXs).
- ❑ Dispensary.
- ❑ Chaplains tent.
- ❑ Chapel and movie tent.
- ❑ Post-exchange van.
- ❑ Bulletin boards.

❑ Maintenance areas.

- ❑ Maintenance tents.
- ❑ Bulk-POL facilities (storage and issue).
- ❑ Package-POL storage.
- ❑ Waste-oil storage.
- ❑ Tire-changing area.
- ❑ Welding shop.
- ❑ The Army Maintenance Management System (TAMMS) and PLL office.
- ❑ Storage-van locations.

❑ Sanitary and support facilities.

- ❑ Water point.
- ❑ Shower point (with leach field or equivalent).
- ❑ Shaving stands.
- ❑ Sump areas.
- ❑ Latrines.
- ❑ Urinals.
- ❑ Human–waste burning areas.
- ❑ Trash–burning pit.
- ❑ Power supply/generators.
- ❑ Laundry areas.

❑ Dining facility.

- ❑ Food-preparation area.
- ❑ Food-serving area.
- ❑ Eating area (covered or shaded).
- ❑ Refrigerator vans.
- ❑ Dry-goods storage area.

- ❑ Hand-wash point.
- ❑ Grease-disposal sumps.
- ❑ Pot and pan wash area.
- ❑ Water point.
- ❑ Grease pit.
- ❑ Cooks' quarters.

❑ Other facilities.

- ❑ Helipads for MEDEVAC and VIPs.
- ❑ Aviation refuel areas.
- ❑ Motor parks.
- ❑ Sports facilities, such as soccer and softball fields and tennis and volleyball courts.
- ❑ Perimeter road.
- ❑ Perimeter fencing (single or multiple belt).
- ❑ Perimeter lighting.
- ❑ Guard shacks.
- ❑ Guard towers.
- ❑ Flagpoles.
- ❑ Generator enclosure sites for mobile electric supply.
- ❑ Bunkers.
- ❑ Fighting positions.
- ❑ Berms for fuel points.
- ❑ Water towers or platforms for water blivets to provide water pressure.

❑ Construction. Ensure that the base camp construction party takes adequate equipment and SKOs to perform its mission. This may include hand tools, DeWalt saws, pioneer electric–tool trailers, earth–moving equipment, and small emplacement excavators (SEES). Rented trenching machines or backhoes may be required for buried utilities. Do not mow wild grasses unnecessarily-exposed earth generates dust.

Redeployment–and–Recovery-Considerations Checklist. The following considerations apply to planning for redeployment and recovery:

❑ Base-camp breakdown.

- ❑ Breakdown of each tent area.
- ❑ Disassembly of plywood floors.
- ❑ Earthwork to return the site to its original condition.

- ❑ Dates for—
  - ❑ Last hot meal in dining facility.
  - ❑ Last shower.
  - ❑ Last movie.
  - ❑ Last mail delivery.
  - ❑ Last resupply delivery.
- ❑ Closure of Army exchange (PX) van.
- ❑ Closure of S4 yard.
- ❑ MHE requirements.
- ❑ Commercial vans and military-owned remountable containers (MILVANs) positioning.
- ❑ Base-camp support-cntract phaseout.
- ❑ Closure of burn pits and field latrines.
- ❑ Removal of security wire.
- ❑ Electrical-system removal.
- ❑ Turn-in of leased equipment.
- ❑ Maintaining security throughout the base-camp breakdown.
- ❑ Turn-in or storage of base-camp materials, such as tents, plywood tent floors, concertina, wire, and pickets.

❑ Movement.

- ❑ Phased movement of equipment as it is no longer needed.
- ❑ Sensitive item inventory. List the types of items, serial numbers, and how they will be transported (vehicle/ship ID or the person carrying them).
- ❑ Security force for convoys.
- ❑ Staging area for equipment at ports.
- ❑ Initial and final cleaning of equipment. Plan for an initial cleaning in the vicinity of the base camp and a final cleaning in the vicinity of the port. Consider contracting steam cleaners. Repaint unit markings on vehicles and equipment as required.
- ❑ Inspection of equipment before shipment. Identify nonmission-capable equipment and identify needed repair parts.
- ❑ Logistics application of automated marking and reading symbols (LOGMARS) labeling.
- ❑ Customs inspection of equipment, personal gear, and personnel.
- ❑ Numbers of personnel to be moved by airlift.
- ❑ Target dates for each movement.
  - ❑ Advance-party personnel.

    - ❑ Main–body equipment.
    - ❑ Main–body personnel.
    - ❑ Trail–party equipment.
    - ❑ Trail–party personnel.

- ❑ Recovery.
    - ❑ Personnel accountability.
    - ❑ Sensitive-items inventory.
    - ❑ Equipment and SKOs inventories.
    - ❑ Equipment and tool maintenance.
    - ❑ After–action reports and lessons learned.
    - ❑ Homecoming ceremony.
    - ❑ Awards and soldier recognition.
    - ❑ Rail equipment requirements.
    - ❑ Block leave.

Construction-Considerations Checklist. The following considerations apply to construction operations in support of a deployment:

- ❑ Drainage designs. Underdesigned culverts, headwalls, outfalls, and bridge supports may be damaged by seasonal rains. Failure to stabilize roadway shoulders causes loose material along roadways to clog culverts, aggravating drainage problems. Planners should seek and use terrain and climatology studies. Planners should conduct a thorough ground reconnaissance with local HN personnel before beginning construction and ensure that designs meet local requirements.
- ❑ Quarry and rock-crusher operations. Inefficient quarry, rock-crusher, and haul operations may limit the ability of units to get aggregate to project sites. Adequate lead time for quarry and crusher setup must be included in the project management planning. Planners must develop an integrated maintenance program for quarry, rock-crusher, and haul equipment. Planners must also develop an efficient procedure for the transportation of aggregate, including well–planned haul routes. If possible, rock crushers should begin operations early and stockpile material before the beginning of construction. Planners must determine if there are any local restrictions or peculiar hazards associated with blasting operations.
- ❑ Quality control and quality assurance. Units should establish and enforce quality-control and quality–assurance programs to ensure that high standards of construction are maintained. If there are not enough trained, experienced personnel within the unit, request them from the next higher headquarters or the engineer staff of the CINC responsible for the area. Areas of particular concern include earthwork, concrete placement, and drainage structures. Coordination with HN officials will assist in developing construction standards.
- ❑ Construction materials. Construction materials contracted from local sources must be requested as soon as possible. Planners will face long lead times in some areas. The quality of materials must be checked by unit personnel. The quality of materials such as ready–mix concrete, cinder block, and clay pipe may vary from location to location. This may require design adjustment.

NOTE: Country–by-country construction-material availability and compatibility information may be obtained from Huntsville Division, USACE, AFCS Branch, ATTN: ED-SY (AFCS), Post Office Box 1600, Huntsville, Alabama 35807.

- ❑ Construction water. Planners must identify sources of construction water and methods of transporting it to construction sites. Water will be required for compaction, cement-soil stabilization, mixing and curing concrete, mixing mortar, and other tasks identified during construction planning.
- ❑ Surveying. Construction plans and drawings should be verified by the surveyors on the site-survey team. Incorrect initial survey data affects not only elevations on the drawings but also earthwork estimates, fill-material requirements, and so forth. Planners should not assume that the initial survey for a project was done correctly.
- ❑ Engineer construction equipment. Planners must determine if MTOE equipment is adequate to accomplish the construction mission. This applies not only to the quantity of equipment but also the type. For example, soil conditions in a region may not support the ground pressure of tractors organic to a particular unit. It may be necessary for the unit to contract smaller tractors or tractors with swamp tracks. Conditions may dictate the use of tracked backhoes as a substitute for the JD410 or SEE. The site-survey team should obtain this information from HN civilian agencies such as the transportation ministry, the HN military, and local construction contractors.
- ❑ Construction techniques. Basic Army reference manuals for construction primarily apply to construction in temperate climates such as CONUS or Europe. The site-survey team must determine if special construction techniques are required for wet tropical conditions, desert environment, and so forth. They must also determine if there are any HN design requirements or prohibitions. The team can get this information from HN agencies such as the transportation ministry, the HN military, and local construction contractors. Planners can also obtain this information from after–action reports from other units that have operated in the same area.
- ❑ Safeguarding the environment of the HN is a key aspect of construction operations. Planners must integrate environmental considerations in everything associated with the project. This includes quarrying operations, haul operations, vehicle and equipment maintenance, and refueling operations, as well as actual construction.

<u>Topographic-Engineering-Considerations Checklist.</u> The following considerations apply to topographic support for a deployment:

- ❑ Identify and list the topographic and terrain–analysis products required to support the deployment.
- ❑ Determine the availability and adequacy of existing topographic and terrain–analysis products, to include existing standard and nonstandard maps.
- ❑ Determine what topographic-engineering forces are assigned to provide topographic support.
- ❑ Determine how the command will provide the topographic support to meet the commander's overall mission requirements.
- ❑ Determine requirements for topographic engineer and depot support to meet mission requirements.

- ❑ Determine how deployed topographic engineering forces will conduct topographic operations.
- ❑ Determine topographic-product and terrain–analysis–product supply, storage, distribution, and replenishment requirements and procedures.
- ❑ Determine requirements for the provision of topographic logistics and maintenance support.
- ❑ Integrate transportation requirements for topographic engineering units into the movement plan.
- ❑ Delineate the priority of topographic support to supported units in coordination with the intelligence staff.
- ❑ Determine primary and alternate locations of topographic engineering units, and speciify command and support relationships according to FM 5-105.

Safety-Considerations Checklist. The following considerations apply to the safety aspects of a deployment:

- ❑ Analyze the area of operations with respect to safety,
  - ❑ Climate (including the potential for hot– and cold-weather injuries).
  - ❑ Terrain.
  - ❑ Vegetation.
  - ❑ Wildlife.
  - ❑ Potentially hazardous snakes and insects.
  - ❑ Environment; for example, whether it is highly populated, rural, or remote.
  - ❑ Local traffic rules and driving habits.
  - ❑ Anticipated missions.
- ❑ Use a risk management approach to analyze the mission. Begin with deployment from home station and work through the whole operation, to include redeployment and recovery.
  - ❑ Identify potential hazards.
  - ❑ Assess the hazards.
  - ❑ Develop controls to reduce or eliminate the hazards.
  - ❑ Implement the controls.
  - ❑ Supervise to ensure the controls and standards are enforced.
  - ❑ Evaluate and update as necessary.
- ❑ Conduct safety training and briefings to prepare personnel for upcoming missions.
- ❑ Contact a US Army Corps of Engineers (USACE) district office and request copies of Engineer Manual (EM) 385–1–1 to use as a reference.

Disaster–Relief-Considerations Checklist. The following considerations apply to disaster–relief specific deployments:

- ❑ Local requirements and capabilities.
  - ❑ Determine local government requirements and priorities for assistance.
  - ❑ Determine location, capabilities, and status of HN military engineer units.
  - ❑ Determine location, capabilities, and status of local contractors.
  - ❑ Locate engineer materials and facilities.
    - ❑ Lumber yards.
    - ❑ Hardware stores.
    - ❑ Concrete plants.
    - ❑ Asphalt plants.
    - ❑ Quarries.
    - ❑ Engineer equipment.
- ❑ Refugees.
  - ❑ Identify location of refugee concentrations.
  - ❑ Identify facilities available for refugees where they are concentrated.
  - ❑ Determine if refugee centers have been established.
    - ❑ Types of centers.
    - ❑ Locations.
    - ❑ Dates established.
    - ❑ Establishing agencies.
  - ❑ Additional requirements for the centers, such as power, shelter, water, and sanitation.
- ❑ Transportation systems.
  - ❑ Identify location of key—
    - ❑ Evacuation/emergency routes.
    - ❑ Bridges.
    - ❑ Railroads.
    - ❑ Airfields.
    - ❑ Ports.
  - ❑ Determine if the above systems are operational. For nonoperational systems-
    - ❑ Report location.
    - ❑ Identify damage.

- ❑ Estimate time, material, and equipment required to make them operational.

- ❑ Facilities.
  - ❑ Identify location of facilities.
    - ❑ Water systems.
    - ❑ Food storage facilities.
    - ❑ Sanitation systems.
    - ❑ Communications systems.
    - ❑ Medical facilities.
    - ❑ Dams/waterways.
  - ❑ Determine the capacity of the above facilities.
  - ❑ Determine if trained operators are available.
  - ❑ Determine the status of the above facilities. For nonoperational facilities-
    - ❑ Report location.
    - ❑ Identify damage.
    - ❑ Estimate time, material, and equipment required to make them operational.

- ❑ POC.
  - ❑ Determine in-country POC at the US embassy (missions outside the US).
  - ❑ Determine who is coordinating disaster–relief activities locally.
  - ❑ Determine the POC with the local—
    - ❑ Police.
    - ❑ Civil-defense agency.
    - ❑ Public–works department.
    - ❑ Transportation department.
    - ❑ Land resources agency.
    - ❑ Military units.

<u>Family-Support-Group-Considerations Checklist.</u> The following considerations apply to family support groups:

- ❑ Ongoing actions (preparation for deployment).
  - ❑ Provide lines of communication for family members (both with the unit and among families).
    - ❑ Telephone trees.

- ❑ Address rosters.
- ❑ Newsletters.
- ❑ Unit calendars.
- ❑ Unit–update briefings.

❑ Provide support to family members and make them feel part of the unit.

- ❑ Group meetings.
- ❑ Group projects.
- ❑ Contact telephone calls.

❑ Provide a structure that meets the needs of the unit.

- ❑ Group-leadership structure.
- ❑ Social committee.
- ❑ Welcoming committee.
- ❑ Contact persons.
- ❑ Records keeper.

❑ Provide training to members of the support group to make it more effective.

- ❑ Crisis-intervention skills.
- ❑ Basic communications and listening.
- ❑ Problem solving.
- ❑ Principles of information and referral.
- ❑ Baseline knowledge of community services.
- ❑ Special issues, such as dealing with stress, loneliness, and a death in the family.

❑ Predeployment actions.

❑ Predeployment briefings should address—

- ❑ Description of the exercise.
- ❑ Exercise dates.
- ❑ Support available to families.
- ❑ Telephone tree/chain of concern.
- ❑ Methods for corresponding with deployed spouse.
- ❑ Methods for contacting deployed spouse in an emergency.
- ❑ Methods for picking up mail, checks, and so forth.
- ❑ Current emergency phone numbers.

- ❑ Family emergency information packets should contain—
  - ❑ Telephone numbers.
    - ❑ Personal (friends and family).
    - ❑ Business (military and civilian).
    - ❑ Family–support group.
  - ❑ Important family documents.
    - ❑ Bank book/checkbook/credit cards.
    - ❑ Powers of attorney.
    - ❑ Passports (if overseas).
    - ❑ Location of wills.
    - ❑ Location of insurance policies.
    - ❑ Location of birth certificates.
    - ❑ Location of mutual funds and certificates of deposit.
  - ❑ Information on emergency assistance agencies.
    - ❑ Military/civilian police.
    - ❑ Army Community Service.
    - ❑ Army Emergency Relief.
    - ❑ Red Cross.
    - ❑ Chaplain.
  - ❑ Medical information.
    - ❑ Number for emergency medical care.
    - ❑ Number for family doctor.
    - ❑ First-aid information.
    - ❑ Civilian Health and Medical Program of the Uniformed Services (CHAMPUS) information.
    - ❑ Defense Enrollment Eligibility Reporting System (DEERS) information.
  - ❑ Information on routine services.
    - ❑ Communication with the deployed soldier.
    - ❑ Mail pickup.
    - ❑ Housing problems.
    - ❑ Lost ID cards.
    - ❑ Pay problems, such as lost checks, no check, and allotment problems.

  - ❑ Legal assistance.
  - ❑ Auto craft shop.

- ❑ Actions during deployment.

  - ❑ Newsletters.
  - ❑ Group meetings.
  - ❑ Group social functions, such as dinners and recreational trips.
  - ❑ Group projects, such as pen pals and volunteer work.
  - ❑ Periodic contact telephone calls or visits.
  - ❑ Unit-update briefings (rear detachment).
  - ❑ Special services such as babysitting.

Appendix C

Engineer Cellular Teams

PURPOSE

This appendix describes various types of engineer cellular teams available to augment other engineer organizations or to be used as separate elements. When operating independently, these teams require significant augmentation of personnel, such as mechanics, medics, and communications personnel; and equipment, such as ambulances, high frequency (HF) radios or tactical satellite (TACSAT) terminals, and administrative vehicles. This information will help planners choose and organize engineer support for missions in environments short of war. These are teams for which tables of organization and equipment (TOEs) currently exist-Army manning levels may limit team availability.

GENERAL

This appendix groups engineer cellular teams by function. For the following types of teams, this appendix outlines the structure, mission, support requirements, allocation, potential use, and references:

- Engineer administrative and headquarters teams.
- Fire-fighting teams.
- Equipment-operating teams.
- Engineer support teams.
- Engineer topographic teams.
- Engineer dredge teams.
- Engineer bridge teams.

ENGINEER ADMINISTRATIVE AND HEADQUARTERS TEAMS

Battalion Headquarters Team. The following information applies to the battalion headquarters team:

- Structure and Equipment. Is organized into seven sections (see Figure C-1, page C-2) and contains 8 officers, 18 noncommissioned officers (NCOs), and 15 other enlisted soldiers. Aggregate strength is 41. Significant equipment includes–

  - Four 3/4–ton utility trucks.
  - Seven 5/4-ton cargo trucks.

- Two 2 1/2–ton cargo trucks.
- One 5–ton cargo truck.
- One tank and pump unit.
- One battalion drafting kit.
- One engineer, platoon carpenter's tool kit.
- Two general mechanic's tool kits.

- Mission. Provides command, control, and administrative support for separate engineer companies and engineer teams. This team can command and control three to seven engineer units.
- Support Requirements. Depends on the unit to which it is attached for supply, food, health, religious, financial, legal, and administrative services, along with unit maintenance.
- Allocation. Provided as require for the command and control of composite battalion organizations at corps and echelons above corps.
- Potential Use. Can provide command and control of engineer assets to support deployed United States (US) forces. It can also provide training for indigenous forces and host-nation (HN) personnel in planning and controlling engineer operations.
- References. TOE-05500LA00 and Field Manuals (FMs) 100-5, 5–100, and 5-116.

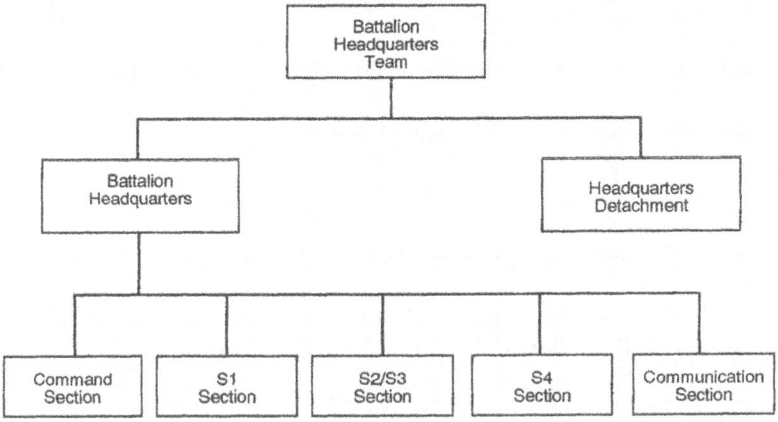

Figure C–1. Battalion–headquarters–team organization.

FIRE-FIGHTING TEAMS

Fire-Fighting Headquarters Team. The following information applies to the fire-fighting headquarters:

- Structure and Equipment. Is a headquarters section containing one officer, two NCOs, and one other enlisted soldier. Significant equipment consists of one aircraft forced-entry–and–rescue equipment set and two 1/4-ton cargo trucks.
- Mission. Controls four to five fire-fighting teams. It also plans fire-fighting programs and area fire protection.
- Support Requirements. Depends on the unit to which it is attached for supply, find, health, religious, financial, legal, and administrative services, along with unit maintenance.
- Allocation. One per four to five fire-fighting teams.
- Potential Use. Can control assets providing base support for deployed US forces. It can also provide training for indigenous forces and for HN civilian fire-fighting personnel.
- References. TOE–05510LA00, Technical Manual (TM) 5-315, FMs 5–116 and 5–100, and Department of the Army (DA) Pamphlet 420–2.

Fire-Truck Team. The following information applies to the fire-truck team:

- Structure and Equipment. Is structured around a fire truck and must be augmented with a water-truck team. The team consists of one NCO and two other enlisted soldiers. Significant equipment consists of one multipurpose, truck-mounted fire-fighting equipment set. This team should be controlled by a fire-fighting headquarters team.
- Mission. Provides fire protection, administers timely and adequate first aid, and implements a fire-prevention program for major facilities: ports or pier sites; petroleum, oils, and lubricants (POL) tank farms and distribution sites; open and closed warehouse facilities or general depots (to include ammunition supply points); fixed or deployed hospitals; and enemy prisoner of war (EPW) and civilian internee camps.
- Support Requirements. Depends on the unit to which it is attached for supply, find, health, religious, financial, legal, and administrative services, along with unit maintenance.
- Allocation. One per major port or pier; one per POL tank farm or distribution center; one per open-storage warehouse facility; two per facility of more than five square miles in size; one per field, station, or general hospital within logic regions 3 through 5; and one per EPW and civilian internee camp.
- Potential Use. Can provide support for deployed US forces. It can also assist HN fire-fighting forces and implement or augment the training program developed by a fire-fighting headquarters team.
- References. TOE-05510LB00, TM 5-315, and FMs 5–116 and 5–100.

Water-Truck Team. The following information applies to the water-truck team:

- Structure and Equipment. Is designed around a water truck. It will augment other fire-fighting teams: one water truck per fire-truck team, or two water trucks per brush–fire truck or crash–truck team. This team consists of two enlisted soldiers. It should be controlled by a fire-fighting headquarters team. Significant equipment consists of one 6,000–gallon, trailer–mounted water distributor and one 5–ton tractor.

- Mission. Provides 6,000 gallons of water (per trip) for fire-fighting purposes. Team members may be used as fire fighters.
- Support Requirements. Depends on the unit to which it is attached for supply, find, health, religious, financial, legal, and administrative services, along with unit maintenance.
- Allocation. One per fire-truck team and two per brush-fire or crash–rescue truck team.
- Potential Use. Can provide support for deployed US forces. It can also assist HN forces and implement or augment the training program developed by a fire-fighting headquarters team.
- References. TOE-05510LC00, TM 5-315, and FMs 5–116 and 5–100.

Brush-Fire-Truck Team. The following information applies to the brush-fire truck team:

- Structure and Equipment. Is structured around a brush-fire truck and must be augmented by two water-truck teams. This team consists of one NCO and one other enlisted soldier. It should be controlled by a fire-fighting headquarters team. Significant equipment includes one 1,000–gallon fire-fighting truck and one multipurpose, truck-mounted fire-fighting equipment set.
- Mission. Can provide protection against and the suppression of grass or brush fires within its area of responsibility. When augmented with additional personnel and tools, the team may be used to fight structural fires on a limited basis.
- Support Requirements. Depends on the unit to which it is attached for supply, food, health, religious, financial, legal, and administrative services, along with unit maintenance.
- Allocation. One per headquarters and headquarters company (HHC), theater army; one per corps support group; one per area support group; one per HHC, corps; and one per HHC, division.
- Potential Use. Can provide support for deployed US forces. It can also assist HN forces and implement or augment the training program developed by a fire-fighting headquarters team.
- References. TOE-05510LD00, TM 5-315, and FMs 5–116 and 5–100.

Aviation Fire and Rescue Team. The following information applies to the aviation fire and rescue team:

- Structure and Equipment. Is structured around a crash–rescue truck and must be augmented by two water–truck teams. This team consists of one NCO and one other enlisted soldier. Significant equipment includes one aircraft forced-entry-and-rescue equipment set, one 3/4--ton utility truck, and one multipurpose, truck-mounted fire-fighting equipment set.
- Mission. Fights aviation fires, extricates personnel and equipment from crashed aircraft, and performs first aid on injured personnel.
- Support Requirements. Depends on the unit to which it is attached for supply, food, health, religious, financial, legal, and administrative services, along with unit maintenance.

- Allocation. Three per HHC, division; one per HHC, separate brigade; and one per headquarters and headquarters troop (HHT), armored tawdry regiment (ACR).
- Potential Use. Can provide support to deployed US forces. It can also assist HN forces and implement or augment the training programs developed by a fire-fighting headquarters team.
- References. TOE-05510LE00, TM 5-315, and FMs 5–116 and 5–100.

EQUIPMENT-OPERATING TEAMS

Forestry Team. The following informatiion applies to the forestry team:

- Structure and Equipment. Is organized in three sections (see Figure C-2, page C-6) and consists of 1 officer, 16 NCOs, and 28 other enlisted soldiers. Aggregate strength is 45. Significant equipment includes-
  - One 20–ton wheel–mounted crane.
  - Two chain saws.
  - One forestry-company supplementary equipment set.
  - One full-tracked, low-speed, medium draw–bar–pull (DBP) tractor.
  - One 9-ton bolster trailer.
  - One 10,000–pound forklift.
  - One trailer–mounted welding shop.
  - One 250 cubic feet per minute (cfm) pneumatic tool and compressor outfit.
  - One 40–ton trailer.
  - One 5–ton bolster truck.
  - One 10–ton tractor.
  - One 5–ton tractor.
  - One trailer–mounted sawmill.
  - One 2 1/2–ton cargo truck.
  - One 5/4–ton cargo truck.
  - Four general mechanic's tool kits.
  - One 1 1/2–ton trailer.
- Mission. Conducts logging and sawmill operations. It also produces 15,000 board feet of rough lumber and timber pilings per day.
- Support Requirements. Depends on the unit to which it is attached for supply, food, religious, health, financial, legal, and administrative services.
- Allocation. As required. It is normally attached to a supply and service battalion or to an engineer group.
- Potential Use. Can provide lumber for deployed US forces. It can also assist HN personnel with civic action projects and provide training for HN personnel. The team may support base development or combat and sustainment engineering operations.
- References. TOE-05520LA00 and FMs 5-342,5-100, 5–116, and 5–104.

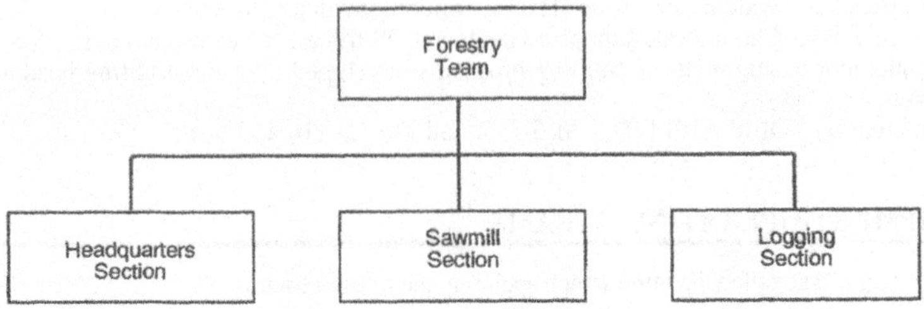

Figure C–2. Forestry–team organization.

Quarry Team, 75 tons per hour (tph). The following information applies to the 75–tph quarry team:

- Structure and Equipment. Is organized into three sections (see Figure C-3) and consists of 12 NCOs and 26 other enlisted soldiers. Aggregate strength is 38. Significant equipment includes-
  - Two full–tracked, low–speed, heavy DBP tractors.
  - One 5–ton cargo truck.
  - Two 750-cfm air compressors.
  - One crushing–and-screening plant.
  - One 4 1/2-cubic-yard loader.
  - Two rock-drilling equipment sets.
  - Four general mechanic's tool kits.
  - One engineer, squad carpenter's tool kit.
  - One engineer, platoon pioneer tool kit.
  - One 5/4–ton cargo truck.
  - Two 15–ton dump trucks.
  - One 12 1/2–ton crane.
  - Three demolition sets.
  - Two 600-cfm pneumatic–tool–and-compressor outfits.
  - One 10–ton tractor.
  - One washing–and-screening plant.
  - One 10–ton trailer.
  - One 600–gallon liquid storage tank.
  - One 40–ton trailer.
  - Two floodlight sets.
  - Four conveyor belts.
  - One aggregate storage bin.
- Mission. Provides qualified personnel and equipment for 24-hour operation of a quarry and rock-crushing plant that provides 75 tons of rock per hour. Equipment is capable of hauling 30 tons of rock per trip.
- Support Requirements. Depends on the unit to which it is attached for supply, food, religious, health, legal, financial, and administrative services.

- Allocation. As required.
- Potential Use. Can provide support for deployed US forces. It can also assist HN personnel with civic action projects and provide training for HN personnel. Support may include base development and maintenance, lines of communication (LOC) development and maintenance, and logistical–facility development and maintenance. Civic action and humanitarian projects may include public–works development, along with road and bridge construction and maintenance.
- References. TOE-05520LC00; TMs 5-332 and 5-330; and FMs 5–100, 5–116, and 5–104.

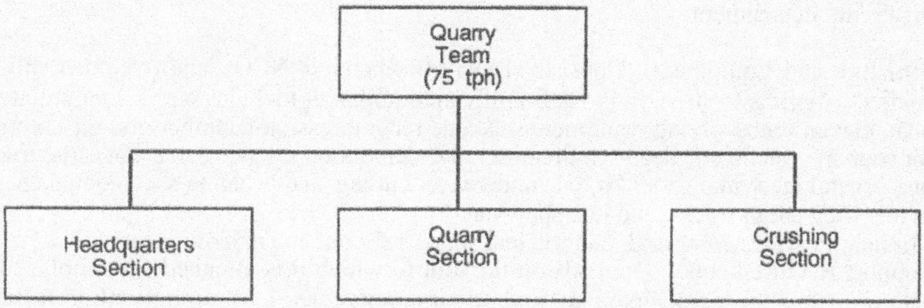

Figure C–3. Quarry–team (75 tph) organization.

Well-Drilling Team, 1,500 feet. The following information applies to the 1,500-foot well-drilling team:

- Structure and Equipment. Consists of four NCOs and five other enlisted soldiers. Aggregate strength is 9. Significant equipment includes-
  - One truck-mounted well-drilling machine (1,500 feet).
  - One 125 gallons per minute (GPM) pump.
  - One 5/4–ton cargo truck.
  - Four fabric tanks.
  - One master mechanic tool kit.
  - Two pipe-fitter's tool kits.
  - One 2 1/2–ton cargo truck.
  - One 5–ton cargo truck.
  - One floodlight set.
- Mission. Provides qualified personnel and equipment for drilling and developing water wells. Equipment should be capable of drilling two wells to a maximum depth of 1,500 feet.
- Support Requirements. Depends on the unit to which it is attached for supply, food, religious, health, financial, legal, and administrative services. Additionally, this team requires support in well-site preparation; resupply and transporting of drilling materials and supplies; water-quality testing, external security; and assistance in concrete, electrical, and welding operations.

- Allocation. As required.
- Potential Use. Can provide support to deployed US forces. It can also assist HN personnel with civic action projects and provide training for HN personnel. The team can also provide support during disaster–relief operations.
- References. TOE-05520LE00 and FMs 5–100, 5–104, 5–116, and 5–166.

ENGINEER SUPPORT TEAMS

Control and Support Diving Detachment. The following information applies to the control and support diving detachment:

- Structure and Equipment. Consists of two officers, eight NCOs, and five other enlisted soldiers. Aggregate strength is 15. Significant equipment includes one 7–man inflatable boat, eleven scuba-diving equipment sets, one recompression-chamber-and-air-compressor support system, one deep-sea-divingset, one demolition set, two 2 1/2–ton cargo trucks, one general mechanic's tool kit, one underwater cutting–and–welding shop equipment set, two 5/4-ton cargo trucks, and one shop van.
- Mission. Provides command, control, and support for one to six diving teams,
- Support Requirements. Depends on the unit to which it is attached for supply, food, religious, health, legal, financial, and administrative services, along with unit maintenance.
- Allocation. One per engineer command (ENCOM) for control of up to six lightweight-diving teams.
- Potential Use. Can provide diving support to deployed US forces and HN personnel. It can also provide assistance and advice on harbor and port clearance, development, and maintenance. The team can also support vessel damage control, maintenance, and repair; offshore petroleum distributions systems; and logistics-over-the-shore (LOTS) operations. It can provide HN forces with dive training and assistance in civic action projects and provide support during river-crossing operations.
- References. TOE-05530LA00 and FMs 5-100, 5-101, 5–116, and 20–11-1.

Lightweight-Diving Team. The following information applies to the lightweight-diving team:

- Structure and Equipment. Is organized into two sections (see Figure C-4) and consists of 1 officer, 10 NCOs, and 6 other enlisted soldiers. Aggregate strength is 17. Significant equipment includes—
  - One 15–man inflatable boat.
  - Two 7–man inflatable boats.
  - Seventeen scuba-diving equipment sets.
  - One recompression-chamber-and-air-support system.
  - One lightweight-diving equipment set.
  - One photographic-support set.
  - Two demolition sets.
  - Two 5/4-ton cargo trucks.
  - One underwater cutting–and–welding shop equipment set.
  - Two outboard motors.
  - One general mechanic's tool kit.

- Mission. Performs scuba, lightweight, or deep-sea diving to a maximum depth of 190 feet in support of light salvage and harbor clearance, repair, and rehabilitation.
- Support Requirements. Depends on the unit to which it is attached for supply, food, religious, health, legal, financial, and administrative services, along with unit maintenance.
- Allocation. One per one to three transportation, floating-craft, or general support (GS) maintenance companies.
- Potential Use. Can provide diving support to deployed US forces and HN personnel. It can provide assistance and advice on harbor and port clearance, development, and maintenance. The team can also support vessel damage control, maintenance, and repair; offshore petroleum distribution systems; and logistics-over-the-shore operations. It can provide HN forces with dive training and assistance in civic action projects and provide support during river-crossing operations.
- References. TOE-05530LC00 and FMs 5–100, 5–101, 5–116, and 20–11–1.

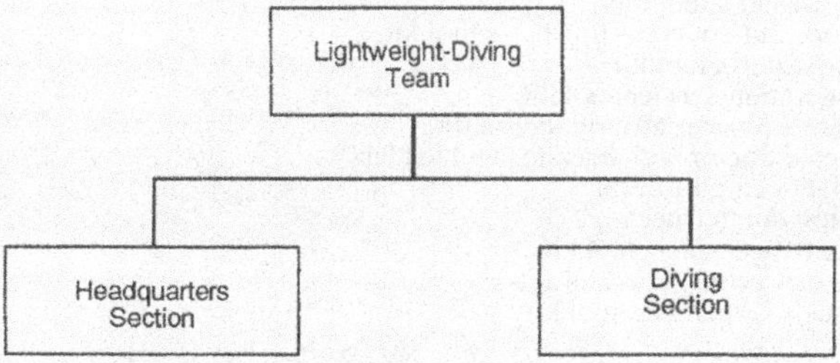

Figure C–4. Lightweight–diving–team organization.

<u>Real Estate Team.</u> The following information applies to the real estate team:

- Structure and Equipment. Contains five officers, five NCOs, and five other enlisted soldiers. Aggregate strength is 15, Significant equipment consists of two 3/4-ton utility trucks, one 5/4-ton cargo truck, and one camera.
- Mission. Performs functions incidental to the acquisition, utilization, and disposal of real property required or occupied by military forces.
- Support Requirements. Depends on the unit to which it is attached for supply, fired, health, religious, financial, legal, and administrative services, along with unit maintenance. It also requires command and control when organized into composite service units.
- Allocation. One per theater army.
- Potential Use, Can provide support for base, logistical–facility, and LOC development for deployed US forces. It can also provide assistance to HN personnel in the execution of real estate operations.
- References. TOE-05530LF00; FMs 5–100, 5–116, and 5–104; and TM 5-300.

Utilities (4,000) Team. The following information applies to the utilities team.

- Structure and Equipment. Is organized into six sections (see Figure C-5) and consists of 2 officers, 15 NCOs, and 43 other enlisted soldiers. Aggregate strength is 60. Significant equipment includes-
  - One road grader.
  - One 2 1/2-cycle loader.
  - One 250-cfm pneumatic–tool–and-com presser outfit.
  - One self–propelled, vibratory roller.
  - Two chain saws.
  - One concrete truck.
  - Three telephone-maintenance trucks.
  - One telephone-construction truck.
  - Two 1/4-ton utility trucks.
  - Three engineer, platoon carpenter's tool kits.
  - Three engineer, squad carpenter's tool kits.
  - One battalion drafting set.
  - Three electrician's tool kits.
  - One mason–and-concrete-finisher's tool kit.
  - Three pipe-fitter's tool kit.
  - One refrigeration-service tool kit.
  - One portable, pioneer electric tool outfit.
  - One wheeled tractor with backhoe and loader.
  - Three 5/4–ton cargo trucks.
  - Three 5–ton dump trucks.
  - Two general mechanic's tool kits.
  - Three platoon carpenter's tool kits.
  - Three squads carpenter's tool kits.
- Mission. Maintains facilities on installations with populations between 2,500 and 4,000. The team maintains utilities and furnishes utilities service and repair, including maintenance of environmental equipment. It also provides insect control.
- Support Requirements. Depends on the unit to which it is attached for supply, food, health, religious, financial, legal, and administrative services. It also requires command and control when organized into composite units.
- Allocation. One per installation with a population of 4,000 individuals. Additional teams will be provided for installations over 4,000. The team is normally attached to an engineer brigade or group.
- Potential Use. Can support base, logistical-facility, and LOC development, operations, and maintenance for deployed US forces. It can also provide HN personnel with construction-skill training and assistance in civic action projects.
- References. TOE-05530LH00; FMs 5-100, 5–116, and 5–104; and TMs 5–610 and 5-683.

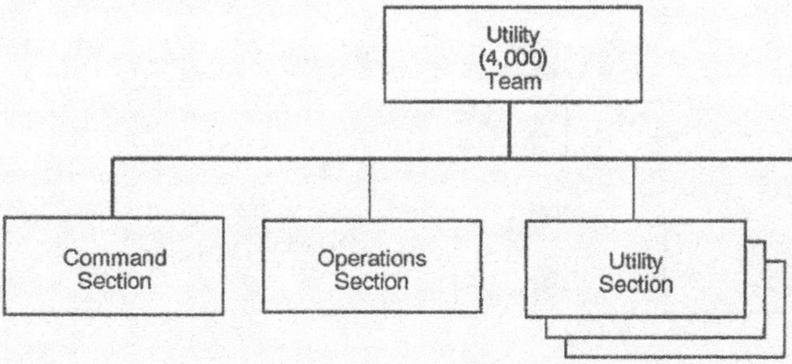

Figure C–5. Utility (4,000) team organization.

Power-Plant-Operation-and Maintenance (O&M) Team. The following information applies to the power–plant-operation-and-maintenance team:

- Structure and Equipment. Is organized into two sections (see Figure C-6, page C-12) and consists of 1 officer and 15 NCOs. Aggregate strength is 16. Significant equipment includes—
  - One semitrailer–mounted electrical–repair shop.
  - One power–plant–maintenance service kit.
  - One electrical–repair–shop equipment set.
  - Four general mechanic's tool kits.
  - One engineer squad tool kit.
  - Three electrician's tool kits.
  - Two lineman's tool kits.
  - One 5/4–ton cargo truck.
  - One 5–ton tractor.
- Mission. Operates and maintains one Army electric plant (500 kilowatts (kw) to 4.5 megawatts (MW)). It also assist in the installation of the electric plant.
- Support Requirements. Depends on the unit to which it is attached for supply, food service, religious, financial, legal, and administrative services, along with unit maintenance. It also requires material-handling equipment for setup, tear down, and day–to-day operations.
- Allocation. One per electric-power-generation plant.
- Potential Use. Can provide general power in support of contingency operations, disaster relief, and civic action projects worldwide. It can support base operations for deployed US forces. It can also provide HN personnel with power–plant operations and maintenance training.
- References. TOE-05530LJ00, FMs 5–100 and 5–104, and TM 5-684.

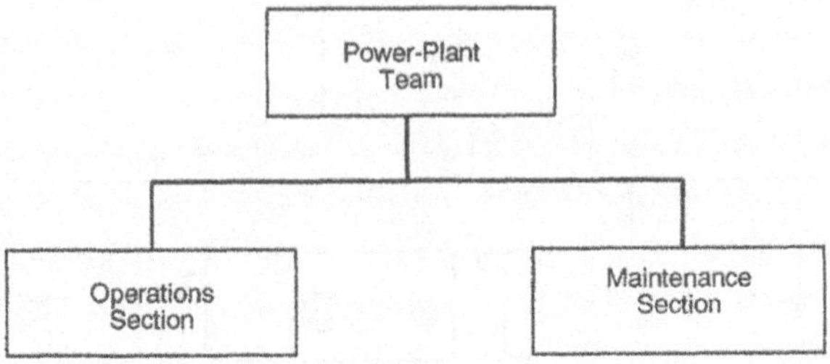

Figure C–6. Power–plant–O&M–team organization.

Power-Line Team. The following information applies to the power-line team:

- Structure and Equipment. Consists of one officer, seven NCOs, and nine other enlisted soldiers. Aggregate strength is 17. Significant equipment includes—

  - One power–line-maintenance service kit.
  - Three 5/4–ton telephone maintenance trucks.
  - One utility construction and maintenance truck.
  - Fifteen lineman's tool kits.
  - One railway electric–power–transmission tool set.
  - One 2 1/2-ton cargo truck.

- Mission. Can construct and maintain up to 60 miles of high–voltage electric power lines.
- Support Requirements. Depends on the unit to which it is attached for supply, food, religious, financial, health, legal, and administrative services, along with unit maintenance.
- Allocation. One per two electric–power–generation plants of 500 kw and above.
- Potential Use. Can provide base, LOC, and logistical-facility support for deployed US forces. It can also provide HN personnel with power–line training and assistance in civic action projects.
- References. TOE-05530LK00, FMs 5–100 and 5–104, and TM 5–684.

ENGINEER TOPOGRAPHIC TEAMS

Topographic-Planning-and-Control Team. The following information applies to the topographic-planning-and-control team:

- Structure and Equipment. Consists of four officers, five NCOs, and two other enlisted soldiers. Aggregate strength is 11. Significant equipment includes one photograph-interpretation kit, one l/4-ton cargo truck, one 3/4-ton utility truck, and one 2 1/2–ton cargo truck.

- Mission. Plans and coordinates the activities of units and agencies in the production and supply of military geographic information and topographic products to the theater army.
- Support Requirements. Depends on the unit to which it is attached for supply, food, religious, health, financial, legal, and administrative services, along with unit maintenance.
- Allocation. One per theater army when no engineer command is assigned.
- Potential Use. Can provide engineer topographic and intelligence support to deployed US forces. It can also provide HN personnel with topographic-planning-and-control training and assistance in civic action projects.
- References. TOE-05540LA00 and FMs 5–100, 5–116, 5-33, and 5–105.

Survey Team. The following information applies to the survey team:

- Structure and Equipment. Consists of 7 NCOs and 19 other enlisted soldiers. Aggregate strength is 26. Significant equipment includes one topographic-support set, two infrared–survey–instrument distance-measuring kits, four microwave-survey–instrument distance-measuring kits, and five 5/4-ton cargo trucks.
- Mission. Performs second, third, and fourth order of topographic and artillery-fire-control-support surveys.
- Support Requirements. Depends on the unit to which it is attached for supply, food, religious, health, financial, legal, and administrative services, along with unit maintenance.
- Allocation. As required, not more than one per corps.
- Potential Use. Can provide engineer topographic and intelligence support to deployed US forces. It can also provide HN personnel with topographic-survey training and assistance in civic action projects.
- References. TOE-05540LB00 and FMs 5–100, 5–116, 5-33, and 5–105.

Terrain–Analysis Squad. The following information applies to the terrain–analysis squad:

- Structure and Equipment. Consists of two NCOs and four other enlisted soldiers. Significant equipment includes one battalion drafting set and one 5/4-ton cargo truck.
- Mission. Produces terrain intelligence for the supported unit (collects, evaluates, and disseminates terrain data and analyzes the effects of terrain on military operations).
- Support Requirements. Depends on the unit to which it is attached for supply, food, religious, health, financial, legal, and administrative services, along with unit maintenance.
- Allocation. One per division or separate brigade as required, not to exceed four at corps or theater level.
- Potential Use. Can provide engineer topographic and intelligence support to deployed US forces. The team can also provide topographic-intelligence training to HN personnel.
- References. TOE-05540LF00 and FM 5–100, 5–116, 5-33, and 5–105.

Printing Squad. The following information applies to the printing squad:

- Structure and Equipment. Consists of four NCOs and four other enlisted soldiers. Significant equipment includes one semitrailer–mounted plate-process section, two semitrailer–mounted press sections, two 5–ton tractors, one general mechanic's tool kit, and one 2 1/2–ton cargo truck.

- Mission. Produces all nonphotographic topographic products.
- Support Requirements. Depends on the unit to which it is attached for supply, food, religious, health, financial, legal, and administrative services, along with unit maintenance.
- Allocation. As required, not to exceed one per corps and one per theater.
- Potential Use. Can provide engineer topographic and intelligence support to deployed US forces. It can also provide HN personnel with map-reproduction training and assistance in civic action projects.
- References. TOE-05540LG00 and FMs 5–100, 5–116, and 5–105.

Technical-Control Section. The following information applies to the topographic-technical–control section:

Structure and Equipment. Consists of one officer, five NCOs, and five other enlisted soldiers, Aggregate strength is 11. Significant equipment includes–

- One topographic-support set, analysis (see note).
- One topographic-support set, collection.
- One topographic-support set, distribution.
- One topographic-support set, synthesis.
- One topographic-support set, information.
- One topographic-support set, operations.
- One topographic-support set, storage.
- Three 5–ton tractors.
- Two 5/4-ton cargo trucks.

NOTE: All topographic support sets are semitrailer mounted.

- Mission. Provides technical control of topographic products.
- Support Requirements. Depends on the unit to which it is attached for supply, food, religious, health, financial, legal, and administrative services, along with unit maintenance.
- Allocation, As required, not to exceed one per corps and one per theater not supported by an engineer topographic company.
- Potential Use. Provides engineer topographic and intelligence support to deployed US forces. It also provides HN personnel with topographic-planning-and-control training and assistance in civic action projects.
- References. TOE-05540LH00 and FMs 5–100, 5–116, 5-33, and 5–105.

Terrain Team (Control). The following information applies to the topographic+ontrol terrain team:

- Structure and Equipment. Consists of one officer, one NCO, and one other enlisted soldier. Significant equipment includes one division topographic-support set and one 1/4-ton utility truck.
- Mission. Provides staff advice and assistance to the supported unit for the control of terrain and analysis squads in direct support of a corps, division, or separate brigade.
- Support Requirements. Depends on the unit to which it is attached for supply, food, religious, health, financial, legal, and administrative services, along with unit maintenance.

- Allocation. One per corps, division, and separate brigade.
- Potential Use. Can provide engineer topographic and intelligence support to deployed US forces. It can also provide HN personnel with topographic-planning-and-control training.
- References. TOE-05540LI00 and FMs 5–100, 5–116, 5-33, and 5–105.

Platoon Headquarters. The following information applies to the topographic platton headquarters:

- Structure and Equipment. Consists of one officer, one NCO, and one other enlisted soldier. Significant equipment includes one l/4-ton utility truck.
- Mission. Provides command and control of attached squads.
- Support Requirements. Depends on the unit to which it is attached for supply, food, religious, health, financial, legal, and administrative services, along with unit maintenance.
- Allocation. One per two to four engineer squads not organic to the unit attached.
- Potential Use. Can provide engineer topographic and intelligence support to deployed US forces. It can also provide HN personnel with topographic-planning-and-control training and assistance in civic action projects.
- References. TOE-05540LJ00 and FMs 5–100, 5–116, 5-33, and 5-105.

Topographic-Maintenance Team C. The following information applies to the topographic-maintenance team:

- Structure and Equipment. Consists of seven NCOs and eight other enlisted soldiers. Aggregate strength is 15. Significant equipment includes-

  - Three general mechanic's tool kits.
  - Three office-machine tool kits.
  - Nine refrigerator tool kits.
  - Three 5/4–ton cargo trucks.

- Mission. Provides intermediate-level maintenance for topographic, reproduction, air conditioning, and power–generation equipment.
- Support Requirements. Depends on the unit to which it is attached for supply, food, religious, health, financial, legal, and administrative services, along with unit maintenance.
- Allocation. One per engineer topographic battalion.
- Potential Use. Can provide engineer-topographic maintenance support to deployed US forces. It can also provide HN personnel with topographic-maintenance training and support for civic action projects.
- References. TOE-05540LM00 and FMs 5–100, 5–116, 5-33, and 5–105.

ENGINEER DREDGE TEAMS

Seagoing-Hopper-Dredge Team. The following information applies to the seagoing-hopper-dredge team:

- Structure and Equipment. This team consists of 10 officers, 33 NCOs, and 27 other enlisted soldiers. Aggregate strength is 70. It is organized into two sections a deck-operations section and a marine-engineer section. Significant equipment includes-

  - Two bridge-erection boats.
  - One amphibious tool set.
  - One engineer, squad carpenter's tool kit.
  - One electrician's tool kit.
  - Fifteen rail and marine tool kits.
  - One machinist tool kit.
  - One rigging and wire-rope tool kit.
  - One small–arms tool kit.
  - One welder's tool kit.
  - Three sheet-metal tool kits.

- Mission. Operates and maintains one medium-class seagoing hopper dredge to facilitate the construction, maintenance, and rehabilitation of waterways and ports.
- Support Requirements. Depends on the unit to which it is attached for supply, food, religious, health, legal, financial, and administrative services. It does have limited medical capability with one medic assigned.
- Allocation. As required.
- Potential Use. Can provide waterway clearance and development support to deployed US forces. It can also provide dredge training and assist HN personnel in civic action projects.
- References. TOE-05550LB00; FMs 5–100, 5–116, 5–104, 5-480,55-50, and 55-60.

ENGINEER BRIDGE TEAMS

Light-Tactical-Raft Team. The following information applies to the light-tactical-raft team:

- Structure and Equipment. Consists of five NCOs and nine other enlisted soldiers. Aggregate strength is 14. Significant equipment consists of six floating bridge or raft sections, twenty-four outboard motors, twelve 2 1/2-ton cargo trucks, and six ferry conversion sets.
- Mission. Provides equipment and technical advice to the supported unit in the erection of six military load class (MLC)–12 rafts or ferries, or one 264-foot, MLC-16 bridge.
- Support Requirements. Depends on the unit to which it is attached for supply, find, religious, health, legal, financial, and administrative services.
- Allocation. As required. It is normally attached to the engineer unit constructing the rafts or a bridge.
- Potential Use. Can provide river-crossing support for deployed US forces and HN personnel. It can also provide HN personnel with bridge training and assistance in humanitarian projects.
- References. TOE-05580LA00 and FMs 5-100, 5–101, 5–104, and 90–13.

Foot-Bridge Team. The following information applies to the foot-bridge team:

- Structure and Equipment. Consists of one NCO and two other enlisted soldiers. Significant equipment includes the 474-fret floating foot bridge, two 2 l/2–ton cargo trucks, and two 4-ton bolster trailers.
- Mission. Provides equipment and technical advice to the supported unit in the erection of the 474-foot floating foot bridge.
- Support Requirements. Depends on the unit to which it is attached for supply, food, religious, health, legal, financial, and administrative services.
- Allocation. As required. It is normally attached to the engineer unit constructing the bridge.
- Potential Use. Can provide river-crossing support for deployed US forces and HN personnel. It can also provide HN personnel with bridge training and assistance in humanitarian projects.
- References. TOE-05580LB00 and FMs 5–100, 5–101, 5–104, and 90–13.

Assault-Boat Team. The following information applies to the assault–boat team:

- Structure and Equipment. Consists of one NCO and four other enlisted soldiers. Significant equipment includes eighty 15–man boats, four outboard motors, four 2 l/2–ton cargo trucks, and two 1 l/2–ton cargo trailers.
- Mission. Provides equipment and technical advice to the supported unit in assault-river–crossing operations.
- Support Requirements. Depends on the unit to which it is attached for supply, food, religious, health, legal, financial, and administrative services.
- Allocation. As required. It is normally attached to the engineer unit supporting the river crossing.
- Potential Uses. Can provide river crossing support to deployed US forces and HN personnel. It can also provide HN personnel with bridge training.
- References. TOE-05580LC00 and FMs 5–100, 5–101, 5–104, and 90–13.

Appendix D

US Army Corps of Engineers

PURPOSE

This appendix outlines the structure, mission, and functions of the US Army Corps of Engineers (USACE). It provides general information for engineer commanders and staffs to use when interacting with or requesting support from USACE.

GENERAL

USACE performs military construction, environmental-restoration, installation-support, real estate, research-anddevelopment, and civil-works missions in peacetime. It provides a base for conversion of its resources to support war and other national emergency conditions. USACE provides engineering and related services in four broad areas. These are-

- Military construction and support.
- Engineering research and development.
- Water and natural-resource management.
- Support to other government agencies.

USACE is the primary agency that the Federal Emergency Management Agency (FEMA) calls upon to protect life and property during and after natural disasters. Military engineers may interact with USACE in any of a number of ways. An engineer unit performing a construction mission may receive technical assistance from a USACE laboratory. Engineers involved in a contingency operation (CONOP) may support or be supported by a USACE prime power team. Personnel from an engineer district may provide design or construction expertise to units that have limited experience internally. For example, USACE may send personnel to assist a unit building roads in Latin America. The normal linkage between engineer troop units and USACE is at the district level. (Note: Within this appendix, documents listed in parenthesis provide additional information on specific programs or activities being addressed.)

ORGANIZATION

The major Army command (MACOM) USACE consists of—

- Headquarters, US Army Corps of Engineers (HQUSACE).
- Thirteen major subordinate commands (engineer divisions).
- Forty engineer districts.

- Four research and development laboratories.
- Other field operating agencies (FOAs).

It also includes any troop units that maybe assigned to USACE.

HQUSACE serves as the MACOM headquarters as follows:

- The commanding general (CG), USACE, serves in a dual capacity. He also acts as the Chief of Engineers (COE). The COE is an Army Staff position (Army Regulation (AR) 10–5). The COE is the principal adviser in Headquarters, Department of the Army (DA), for all engineering and environmental matters.
- The Secretary of the Army has made the CG, USACE, the executive agent for several Department of Defense (DOD) areas of responsibility. Elements of the USACE headquarters staff or subordinate FOAs carry out these responsibilities. These elements coordinate with DA staff elements and the Office of the Chief of Engineers.
- The Assistant Chief of Engineers (ACE) on the Army Staff performs all DA–level duties of the COE. He interacts with the DA Staff on executive-agent responsibilities that USACE performs for the COE.
- Civil–works missions and staff remain part of the USACE headquarters and field structure. This reduces administrative and overhead staffing for military and civil–works programs.

The engineer divisions and districts perform—

- Military construction.
- Environmental-program support.
- Installation support.
- Civil works.
- Real estate functions.

The following laboratories and other separate FOAs are under direct HQUSACE supervision. They perform engineering research and development and provide specialized engineering and environmental support to other USACE elements and to other Army commands and activities. These laboratories and activities include:

- <u>United State (US) Army Topographic Engineering Center (TEC), Humphreys Engineer Center, Virginia.</u> TEC performs a full range of research and development services, from basic research through engineering development in the topographic sciences. TEC is responsible for delivering high–precision positioning systems for the field artillery. The center provides target positioning equipment for tactical missile units. TEC also develops systems for reproducing maps portraying specific military geographic information needed by the Army in the field.
- <u>US Army Engineer Waterways Experiment Station (WES), Vicksburg, Mississippi.</u> WES was originally established in 1929 to develop comprehensive plans for flood control in the Lower Mississippi Valley. The lab performs basic and applied research in broad fields including—

  - Hydraulics.
  - Coastal engineering.
  - Soil and rock mechanics.

- Earthquake engineering.
- Concrete.
- Expedient construction.
- Vehicle mobility.
- Environmental relationships.
- Pavements.
- Protective structures.

Within USACE, WES is also responsible for research and development in the areas of—

- Countersurveillance.
- Military hydrology.
- Mine/countermine.
- Stream bank erosion.
- Coastal engineering.
- Water conservation and supply.

- US Army Cold Regions Research and Engineering laboratory (CRREL), Hanover, New Hampshire. CRREL was established in 1961 to improve the winter combat capability of the Army. It was also chartered to study the natural phenomena of cold regions and to devise methods for building, traveling, living, and working there. CRREL performs basic and applied research in support of the following broad mission areas:

  - Winter combat operations.
  - Winter battlefield environment.
  - Cold–regions facilities.
  - Ice engineering.
  - River–ice management.
  - Civil–works remote sensing.
  - Cold–regions hydrology.

- US Army Construction Engineering Research Laboratory (CERL), Champaign, Illinois. Established in 1969, CERL conducts research and development to support Army programs relating to infrastructure and the environment, with emphasis on military construction and base operations and maintenance. A primary goal is to improve construction quality and energy efficiency while still safeguarding the environment.
- US Army Engineering and Housing Support Cnter (USAEHSC),Fort Belvoir, Virginia. USAEHSC is a field operating activity of USACE dedicated to supporting military installations. Its mission is to assist Army Directorates of Engineering and Housing, major commands, and the COE in their efforts to improve the quality of life on military installations. A key responsibility of USAEHSC is the Army Prime Power Program. See Chapter 3 for details.

MISSION

The mission of USACE is to-

- Manage and execute engineering construction and real estate programs for the US Army and Air Force. USACE performs research and development in support of these programs (ARs 70-1 and 71-9).
- Manage and execute installation support programs for Army installations (AR 420–10).
- Manage and execute civil–works programs, including the design, planning, engineering, construction, and research and development functions in support of these programs.
- Manage and provide execution support to the Army Environmental Program. USACE performs research and development in support of this program (AR 200–1).
- Perform research and development through nonsystems-pecific advance development in systems, specialized equipment, procedures, and techniques relevant to engineer support of combat operations (ARs 70–1 and 71–9).
- Develop and maintain a capability to mobilize in response to national-security emergencies, domestic emergencies, and emergency water planning programs.
- Develop technology and design and construct facilities and structures in support of Army space initiatives (AR 10-5).

FUNCTIONS

USACE develops design criteria according to Army force-modernization requirements. It designs and constructs Army facilities in the US and designated areas overseas (ARs 210–21 and 700–127). USACE performs design and construction support for the US Air Force and Army Reserve on assigned projects in the US and for all DOD services in overseas areas designated by DOD. It also establishes (in consultation with the commands involved) contract award, beneficial occupancy, and construction completion dates and provides complete engineering and construction services for a wide variety of projects including environmental restoration. USACE provides design, construction, or other technical services to non–DOD federal agencies under the authority of separate interagency agreements. USACE also provides selected engineering and related services to foreign governments under the auspices of the Department of State, the US Agency for International Development (USAID), and foreign military sales (FMS). Services range from studies of limited scope and duration to complete design and construction programs.

Civil-works functions are performed under the direction of the Assistant Secretary of the Army (Civil Works). These functions include responsibility for all Corps of Engineers activities that use civil–works resources. USACE is responsible for planning, engineering and design, operations and maintenance, research and development, and the supervision and direction of construction required for water–resources development. It also administers certain laws in the US to protect and preserve the navigable waters and related resources, such as wetlands.

In support of its real estate mission, USACE is responsible for establishing, supervising, and carrying out policies and procedures governing the acquisition, management, and disposal of real property under DA control (AR 10–5). It conducts real estate transactions for DA military and civil-works programs and US Air Force elements in the US. USACE performs this function for other DOD and federal agencies on request. The transactions include purchases, sales, leases, and transfers of real property and interests. USACE is responsible for establishing and maintaining the official real estate historical and cartographic library of all real estate used, owned, leased, and disposed of under DA jurisdiction (AR 405–10). USACE also administers programs and funds for space, services, and facilities furnished to it and the Army National Guard outside the National Capital Region (NCR) by the General Services Administration (GSA) (AR 405–1).

Under the Installation Support Program, USACE provides support to installation Directors of Engineering and Housing. It also provides the DA Staff engineer with expertise in acquiring real property in support of force modernization, stationing, mobilization, and other DA initiatives. USACE manages the effective use of the Army's Real Property Inventory. It also manages the planning for orderly development of Army installations.

In addition to the functions listed above, USACE also develops, maintains, and manages the Army Facilities Components System (AFCS). This system provides pre-engineered facilities for theater of operations (TO) planners, logisticians, and constructors. USACE maintains and manages a reserve of nontactical mobile generators and provides the cadre of qualified personnel needed to deploy and operate the generators. When directed, USACE provides the same services for civil disaster-relief operations. It provides the Army field commands and activities, on request, with engineering, maintenance, and operational support and training related to nontactical mobile generators. USACE also provides temporary, emergency (or standby) electrical–power generation, as required. USACE manages and executes the Army's Prime Power Program (AR 700–128). It acts as the DOD executive agent for—

- The Power Reliability Enhancement Program.
- Preservation of Archeological Resources on Public and Indian Lands.
- Cleanup of hazardous contamination on formerly owned DOD properties.

USACE performs basic, exploratory, nonsystems-specific research and development in systems, equipment, procedures, and techniques relevant to the engineering support of military operations and to material development. Research and development meets Army requirements and supports Army civil–works water–resources mission requirements (AR 70–1). Major responsibilities in support of the military research and development mission are-

- Atmospheric, topographic, and terrestrial sciences (ARs 70–1, 70-38, 71–9, and 115–11).
- Engineer support to combat operations (ARs 70–1 and 71–9).
- Energy (ARs 70–1 and 71–9).
- Environmental quality (ARs 70-1, 71–9, and 200–1).
- Environmental compliance, noise abatement at training and readiness installations, and physical restoration of training lands (AR 200-1).
- Military construction (AR 10–5).
- Facilities operations, maintenance, and repair (AR 70–1).
- Military hydrology (ARs 70–1 and 70-38).
- Digital topographic-data review (ARs 70–1 and 71–9).
- Large space structure and space sensor exploitation (AR 10–5).

GLOSSARY

AAFES Army and Air Force Exchange Service

AC Active Component

ACE Assistant Chief of Engineers

ACR armored cavalry regiment

active barrier A barrier that must be manually or automatically deployed in response to a detected threat.

admin administrative

ADT active duty for training

AFCS Army facilities components system

AFM Air Force manual

AFP Air Force pamphlet

AFR Air Force regulation

aggressor Any person seeking to compromise an asset; aggressor categories include protestors, criminals, terrorists, and subversives.

AGR Active Guard Reserve

AIASA Annual Integrated Assessment for Security Assistance

AM amplitude modulation

Annual Integrated Assessment of Security Assistance A report submitted by the US diplomatic mission which, in addition to an assessment of the host country's capabilities, contains recommended and projected levels of security assistance. Also called AIASA.

ANRC American National Red Cross

antiterrorism The range of defensive measures used to reduce the vulnerability of individuals and property to terrorism. Also called AT.

AOAP Army Oil Analysis Program

AOC	Army Operations Center
AOR	area of responsibility
APO	Army Post Office
APOD	aerial port of debarkation
APOE	aerial port of embarkation
AR	Army regulation
ARFOR	Army force
ARNG	Army National Guard
ASG	area support group
ASL	authorized stockage list
AT	antiterrorism
ATC	air traffic control
ATOIC	Antiterrorism Operations and Intelligence Cell
ATTN	attention
AUEL	automated unit equipment list
AUTOVON	automatic voice network
BIFC	Boise Interagency Fire Center
black propaganda	Propaganda that purports to emanate from a source other than the true one.
C	confidential
C^3	command, control, and communications
CA	civil affairs
CAT	civic action team
CERL	Construction Engineering Research Laboratory
cfm	cubic feet per minute

CFS contract field services

CG commanding general

CHAMPUS Civilian Health and Medical Program of the Uniformed Services

CIA Central Intelligence Agency

CINC Commander in Chief

civic action See military civic action.

CJCS Chairman, Joint Chiefs of Staff

clear zone An area immediately surrounding a building that is clear of all visual obstructions such as shrubbery, trees, heating and/or air–conditioning units, or landscape greater than 4 inches in height.

Co company

COE Chief of Engineers

COL colonel

combatting terrorism Actions, including antiterrorism (defensive measures taken to reduce vulnerability to terrorist acts) and counterterrorism (offensive measures taken to prevent, deter, and respond to terrorism) taken to oppose terrorism throughout the entire threat spectrum.

COMSEC communications security

CONEX container express

conflict An armed struggle or clash between organized parties within a nation or between nations in order to achieve limited political or military objectives. While regular forces are often involved, irregular forces frequently predominate. Conflict is often protracted, confined to a restricted geographic area, and constrained in weaponry and the level of violence.

CONOP contingency operation

CONUS continental United States

COPADS commercial–operated parts depot system

COPARS commercial–operated parts retail system

counterdrug operations, DOD support to Support provided by DOD to free the US from the threat of illegal drugs. This support includes providing information to other US government agencies; conducting training for federal, state, and local law enforcement officials; and giving indirect support such as building target ranges, detecting tunnels, and constructing access roads. Support may also include activities such as working with farmers in an HN to help them develop alternatives to producing crops that contribute to the flow of illegal drugs.

counterinsurgency Those military, paramilitary, political, economic, psychological, and civic actions taken by a government to defeat insurgency.

counterterrorism The full range of offensive measures taken to prevent, deter, and respond to terrorism. Also called CT.

country team The executive committee of an embassy, headed by the chief of mission (normally an ambassador), and consisting of the principal representatives of the government departments and agencies present; for example, the DOS, Defense, Treasury, Commerce, and the USIA, USAID, DEA, and CIA.

CRREL Cold Regions Research and Engineering Laboratory

CSS combat service support

CT counterterrorism

DA Department of the Army

D/AACG departure/arrival airfield control group

DBP draw–bar pull

DC district command

DCO disaster control officer

DCSI Deputy Chief of Staff for Intelligence

DCSOPS Deputy Chief of Staff for Operations and Plans

DEA Drug Enforcement Administration

DEERS Defense Enrollment Eligibility Reporting System

Defense Security Assistance Agency The agency responsible for administering and supervising the DOD component of security–assistance planning and programs. Also called DSAA.

DEH Directorate of Engineering and Housing

DEL deployment equipment list

demonstration An attack or show of force on a front where a decision is not sought; made with the aim of deceiving the enemy. It is similar to a feint with the exception that no contact with the enemy is sought.

deterrence The prevention from action by fear of the consequences. Deterrence is s state of mind brought about by the existence of a credible threat of unacceptable counteraction.

developmental assistance program A program designed to assist the internal development strategy of an HN.

DFA Direct Federal Assistance

DFT deployment for training

disaster–relief operations Operations that provide emergency assistance to victims of natural or man–made disasters abroad. These operations are responses to requests for immediate help and rehabilitation from foreign governments or international agencies. These operations may include refugee assistance, food programs, medical treatment and care, or other civilian welfare programs.

DLEA drug–law enforcement agency

DMA Defense Mapping Agency

DOD Department of Defense

DODAAC DOD activity address code

DOE Department of Energy

DOJ Department of Justice

DOMS Directorate of Military Support

DOS Department of State

DSAA Defense Security Assistance Agency

DSR Damage Survey Report

DTIC Defense Technical Information Center

DZ drop zone

Economic Support Fund A program by which economic assistance is provided on a loan or grant basis to selected foreign governments having unique security problems. The funds are used to finance imports of commodities, capital, or technical assistance in accordance with terms of a bilateral agreement; counterpart funds thereby generated may be used as budgetary support. These funds enable a recipient government to devote more of its resources to defense and security purposes, than it could otherwise do, without serious economic or political consequences. Also called ESF.

Economy Act Section 686, Title 31, US Code, which authorizes executive departments independent government establishments to order materials, supplies, equipment, work, or services from each other.

elements of national power The various means by which a nation can achieve its national goals. These include the political, economic, informational, and military elements of national power.

ELINT electronic intelligence

EM engineer manual

ENCOM engineer command

end state The ultimate conditions resulting from a course of events.

EOC emergency operations center

EPA Environmental Protection Agency

EPW enemy prisoner of war

ER engineer regulation

ERC exercise–related construction

et seq (et sequens) and the following one

ETSS Extended Training Service Specialists

exclusive standoff zone A controlled area surrounding a facility in which only service and delivery vehicles are allowed. The perimeter of this area is defined by barriers and is set at a standoff distance sufficient to reduce the blast effects of a vehicle bomb.

exercise–related construction Enduring improvements and structures constructed in support of JCS–directed or –coordinated exercises OCONUS. Also called ERC.

FAA Foreign Assistance Act

facility A building or structure.

FAD	federal active duty
FAD	force/activity designator
FAS	Freely Associated States
FBI	Federal Bureau of Investigation
FCO	Federal Coordinating Officer
f'_c	compressive strength of concrete
Feb	February
FEMA	Federal Emergency Management Agency
FID	foreign internal defense
FIDAF	foreign internal defense augmentation force
FM	field manual
FM	frequency modulation
FMS	foreign military sales
FOA	field operating agency

force protection A security program designed to protect soldiers, civilian employees, family members, and facilities and equipment in all locations and situations. This protection is accomplished through planned and integrated application of combatting terrorism, physical security, operations security, and personal protective services. This protection is supported by counterintelligence and other security programs.

foreign assistance Assistance ranging from the sale of military equipment to donations of food and medical supplies to aid survivors of natural and man-made disasters. US assistance takes three forms—development, humanitarian, and security assistance.

foreign internal defense Participation by civilian and military agencies of a government in any of the action programs taken by another government to free and protect its society from subversion, lawlessness, and insurgency. Also called FID.

foreign internal defense augmentation force A specially trained, area-oriented, partially language-qualified, ready force available to the commander of a unified command for the support of operations in situations short of open hostilities and in limited and general war. Foreign internal defense augmentation force organizations may vary in size and capabilities according to theater requirements. Also called FIDAF.

foreign military sales That portion of US security assistance authorized by the Foreign Assistance Act of 1961, as amended, and the Arms Export Control Act of 1976, as amended. This assistance differs from the IMETP in that the recipient provides reimbursement for defense articles and services rendered. Also called FMS.

FORSCOM United States Army Forces Command

fragment–retention film A polyester film applied to glass to minimize the spread of glass fragments when shattered.

Ft fort

FTS field training services

FTX field training exercise

f_y yield strength of steel

GMC General Motors Corporation

GP general purpose

GPM gallons per minute

grey propaganda Propaganda that does not specifically identify any source.

GS general support

GSA General Services Administration

guerrilla warfare Military and paramilitary operations conducted in enemy–held or hostile territory by irregular, predominantly indigenous forces.

HCA humanitarian and civic assistance

HF high frequency

HHC headquarters and headquarters company

HHT headquarters and headquarters troop

high–risk personnel Personnel whose grade, assignment, symbolic value, or relative isolation make them an attractive or accessible terrorist target.

HN host nation

HNS host–nation support

host country A nation in which representatives or organizations of another state are present because of government invitation and/or international agreement. Also called HN.

HQ headquarters

HQUSACE Headquarters, United States Army Corps of Engineers

humanitarian and civic assistance Assistance specifically authorized by Title 10, US Code, Section 401 and provided in conjunction with military operations. Such assistance is limited to medical, dental, and veterinary care provided in rural areas of a country; construction of rudimentary surface transportation systems; well drilling and construction of basic sanitation facilities; and rudimentary construction and repair of public facilities. Also called HCA.

humanitarian assistance Assistance provided by DOD forces, as directed by appropriate authority, in the aftermath of natural or man–made disasters to help reduce conditions that present a serious threat to life and property. Assistance provided by US forces is limited in scope and duration and is designed to supplement efforts of civilian authorities that have primary responsibility for providing such assistance.

HUMINT human intelligence

ID identification

IDAD internal defense and development

IED improvised explosive devices

IFF identification, friend or foe

IID improvised incendiary devices

IMETP International Military Education and Training Program

IMINT image intelligence

Inc incorporated

installation A post, caserne, compound, and so forth.

insurgency An organized movement aimed at the overthrow of a constituted government through the use of subversion and armed conflict.

internal defense and development The full range of measures taken by a nation to promote its growth strategy and protect itself from subversion, lawlessness, and insurgency. Also called the IDAD strategy.

internal development Actions taken by a nation to promote its growth by building viable institutions (political, military, economic, and social) that respond to the needs of its society.

International Military Education and Training Program Formal or informal instruction provided to foreign military students, units, and forces on a nonreimbursable (grant) basis by offices or employees of the US, contract technicians, and contractors. Instruction may include correspondence courses; technical, educational, or informational publications; and media of all kinds. Also called IMETP.

IPB intelligence preparation of the battlefield

ISO International Standardization Organization

JAG Judge Advocate General

Jan January

JCS Joint Chiefs of Staff

Jnt Pub joint publication

JSAT Joint Security Assistance Training

JTF joint task force

kip (kilopound) 1,000 pounds

ksi kips per square inch

kw kilowatt

LAW light antitank weapon

lb pound

level of protection The degree to which an asset is protected against an aggressor's tactics based on the asset's value. Levels of protection refer to the amount of damage a facility is allowed to sustain or the probability that an aggressor will be defeated by the protective system.

LIC low intensity conflict

LOC lines of communication (logistic routes)

LOGMARS logistics application of automated marking and reading symbols

LOI letter of instruction

LOTS logistics over the shore

low intensity conflict Political–military confrontation between contending states or groups below conventional war and above the routine peaceful competition among states. This confrontation frequently involves protracted struggles of competing principles and ideologies. Low intensity conflict ranges from subversion to the use of armed force. It is waged by a combination of means, employing political, economic, informational, and military instruments. Low intensity conflicts are often localized, generally in the Third World, but contain certain regional and global security implications. Also called LIC.

LSE logistical support element

Ltd limited

MAC Military Airlift Command

MACOM major Army command

major disruptions on installations, units, and facilities Acts, threats, or attempts to to commit such acts as kidnapping, bombings, extortion, hijacking, ambushing, major weapons thefts, arson, assassination, and hostage taking on military installations, unit, or facility. These acts cause widespread publicity requiring special response, tactics, and management.

mandate A commission, authorization, or charter of authority given to a person or organization to carry out specific responsibilities.

Mar March

MARS Military Affiliate Radio System

max maximum

MCA Military Construction, Army

MEDEVAC medical evacuation

MHE materials handling equipment

MILCON military construction

military civic action The use of preponderantly indigenous military forces on projects useful to the local populace at all levels in such fields as education, training, public works, agriculture, transportation, communications, health, sanitation, and others contributing to economic and social development which would also serve to improve the standing of the military forces with the population. (US forces may at times advise or engage in military civic actions in overseas areas.)

military operations short of war Operations conducted in a peacetime environment or under conditions of conflict.

MILVAN military–owned demountable container

min minimum

minimum measures Protective measures applied to all facilities regardless of the identified threat. These measures offer defense or detection opportunities for a small cost, facilitate future upgrades, and deter aggressive acts.

mission–essential vulnerable areas Facilities or activities within the installation that, by virtue of their function, the commander evaluates as areas vital to the successful accomplishment of the installation's mission. This includes areas not essential to the installation's operational mission but which, by nature of the activity, are considered vulnerable to theft, trespass, damage, or other criminal or terrorist activity.

MLC military load class

mm millimeter

MO Missouri

mobile training team A team that consists of one or more US personnel drawn from service resources and sent on temporary duty to a foreign nation to give instruction. The mission of the team is to provide, by training instructor personnel, a military service of the foreign nation with a self–training capability in a particular skill. Also called MTT.

MOGAS motor gasoline

MOI memorandum of instruction

MOS military occupational specialty

MOU Memorandum of Understanding

MOUT military operations on urbanized terrain

mph miles per hour

MSC major subordinate command

MTOE modification table of organization and equipment

MTT mobile training team

MW megawatt

National Command Authorities The President and the Secretary of Defense or their duly deputized alternates or successors. Commonly referred to as NCA.

nation assistance Political, economic, informational, and military cooperation between the US and the government of another nation, with the objective of promoting internal development and the growth of sustainable institutions within that nation. This assistance corrects conditions that cause human suffering and improves the quality of life of the nation's people. Nation assistance is conducted to promote stability within the world, as well as for humanitarian reasons.

NATO North Atlantic Treaty Organization

NBC nuclear, biological, and chemical

NCA National Command Authorities

NCO noncommissioned officer

NCOIC noncommissioned officer in charge

NCP National Contingency Plan

NCR National Capital Region

NCS National Communications System

NEO noncombatant evacuation operation

NOAA National Oceanographic and Atmospheric Administration

noncombatant evacuation Operations that relocate threatened civilian noncombatants from operations locations in a foreign country or HN. These operations normally involve US citizens whose lives are in danger. These operations may also include selected HN natives and third country nationals. Also called NEO.

nonexclusive standoff zone A controlled area used as an extension to an exclusive standoff zone for the high and very high threat severity levels. It encloses an exclusive standoff zone that is established for a medium threat severity level. A nonexclusive standoff zone provides less restrictive land use than an exclusive standoff zone. Its perimeter is defined by barriers and set at a standoff distance sufficient to reduce the effects of a truck bomb.

NRT National Response Team

NWS National Weather Service

NY New York

OC on center

OCONUS outside continental United States

ODCSOPS Office of the Deputy Chief of Staff for Military Operations and Plans

ODT oversea deployment training

OIC officer in charge

O&M operation and maintenance

OPCON operational control

operational continuum A portrayal of the variety of conditions and the range of threat environments in which the US military operates. This continuum consists of three general states: peacetime, conflict, and war. There is no precise distinction between where one state ends and another begins.

operations to restore order A type of contingency operation intended to establish or restore peace and order through the use of force. Also known as peacemaking.

OPFOR opposing forces

OPORD operation order

OPSEC operations security

OSC on–scene coordinator

OST order ship time

PAO public affairs officer

passive barrier A barrier that is permanently deployed and does not require responses to be effective.

PDA Preliminary Damage Assessment

peacekeeping operations Military operations conducted with the consent of the belligerent parties to a conflict to maintain a negotiated truce and to facilitate a diplomatic resolution of a conflict between the belligerents. Also called PKO.

peacemaking operations See operations to restore order.

PEP US Army Personnel Exchange Program

PHS Public Health Service

PKO peacekeeping operation

PL public law

PLL prescribed load list

PMCS preventive maintenance checks and services

PMO Provost Marshal's Office

POC point of contact

POD port of debarkation

POL petroleum, oils, and lubricants

polycarbonate glazing A plastic glazing material with enhanced resistance to ballistic or blast effects. It is also used for forced–entry resistance.

POM preparation for oversea movement (units)

POP performance oriented packaging

POR preparation of replacements for oversea movement

Posse Comitatus Act An act (18 US Code 1385) that prohibits the use of the Army or Air Force to execute any law except in cases and under circumstances expressly authorized by the Constitution or Act of Congress.

POW prisoner of war

predetonation screen A fence or screen that causes antitank rounds to "dud" (become inoperative) or to detonate before reaching its target. When placed at the proper distance for the facility construction, the screen will prevent a facility's exterior from being penetrated by an antitank round.

protective system A system that integrates all the protective measures and procedures required to protect assets against an anticipated threat. The ideal protective system deters, defends against, detects, and defeats aggressors.

psi pounds per square inch

PSYOP psychological operations

PX Army exchange

raid An operation, usually small in scale, involving a swift penetration of a hostile territory to secure information, confuse the enemy, or to destroy his installations. A raid ends with a planned withdrawal on completion of the assigned mission.

RC Reserve Components

rescue-and-recovery operations Operations which may include the rescue of US or friendly foreign nationals or the identification, location, and recovery of sensitive equipment or items critical to US national security.

ROE rules of engagement

RON remain overnight

RPG rocket-propelled grenade

RRT Regional Response Team

R&U repair and utilities

rules of engagement Directives issued by competent military authority that delineate the circumstances and limitations under which US forces will initiate and/or continue combat engagement with other forces encountered. Also called ROE.

S1 Personnel Officer (US Army)

S3 Operations and Training Officer (US Army)

S4 Supply Officer (US Army)

SAD state active duty

SAO security assistance organization

SARSS Standard Army Retail Supply System

SCIF sensitive compartmented information facility

SECNAVINST Secretary of the Navy Instruction

security assistance Group of programs authorized by the Foreign Assistance Act of 1961, as amended, and the Arms Export Control Act of 1976, as amended, or other related statutes by which the the US provides defense articles, military training, and other defense-related services, by grant, credit, or cash sales, in furtherance of national policies and objectives.

security assistance organization All DOD elements located in a foreign country with responsibilities for carrying out security assistance management functions. Examples are military assistance advisory groups, military missions and groups, liaison groups, defense attache personnel, and other groups which perform security assistance functions. Also called SAO.

security assistance surges Accelerated security assistance provided by the US when an allied or friendly nation faces an imminent threat. During these surges, operations usually focus on logistical support.

SEE small emplacement excavator

SF Special Forces

show of force A mission carried out to demonstrate US resolve whereby US forces are deployed to defuse a situation that may be detrimental to US interests or national objectives.

SKO sets, kits, and outfits

SME subject–matter expert

SMEE Subject–Matter Expert Exchanges

SOF special operations forces

SOFA Status of Forces Agreement

SOP standing operating procedure

special threat Any situation involving a sniper, a barricaded criminal, or a hostage taker or any terrorist incident that requires special response and/or reaction, manpower, management, training, and equipment.

SPOD sea port of debarkation

SPOE sea port of embarkation

SSB single side band

SSSC self–service supply centers

St saint

standoff distance A distance maintained between a facility and the potential location for an explosive detonation to reduce the effects of the blast. Standoff distances vary depending on the facility's component construction, the level of protection required, and the explosive's weight.

standoff zone See exclusive and nonexclusive standoff zone.

STARTEX start of the exercise

Status of Forces Agreement Agreements between a HN and a second nation that establish the detailed legal status of the second nation's forces while operating within the HN. Also called SOFA.

strike operations Combat operations in zones under insurgent control or in contested zones. Strike operations are targeted against insurgent tactical forces and bases outside areas of government control. Other internal defense activities may support tactical forces during combat operations. Strike forces normally do not remain in the area of operations after mission accomplishment.

STU secure telephone unit

subversion Action designed to undermine the military, economic, psychological or political strength of a regime.

sustainment engineering The range of activities that provide a deployed force with the construction, LOC maintenance and repair, airfield–damage repair, battle–damage restoration, and minefield clearing needed to sustain operations.

TACCS Tactical Army Combat Service Support (CSS) Computer System

TACSAT tactical satellite

tactical force protection Active and passive measures designed to deter and defeat threats directed toward deployed individuals, units, field sites, and equipment in support of US missions in a LIC environment.

TAFT technical assistance field team

TAMMS The Army Maintenance Management System

TAT technical assistance team

TDA tables of distribution and allowances

TDS Trial Defense Service

TEC Topographic Engineering Center

terms of reference The guidance that governs the implementation of US participation in a PKO; it is published by the designated executive agent for DOD. These terms of reference describe the mission, command relationships, organization, logistics, accounting procedures, coordination and liaison, and responsibilities of the US military units and personnel assigned to or supporting the peacekeeping force.

terrorism The calculated use of violence or the threat of violence to inculcate fear; intended to coerce or intimidate governments or societies in the pursuit of goals that are generally political, religious, or ideological.

THREATCON threat condition

TI technical inspection

TM technical manual

TMDE test, measurement, and diagnostic equipment

TNT trinitrotoluene

TO theater of operations

TOC tactical operations center

TOE table(s) of organization and equipment

tph tons per hour

TRADOC US Army Training and Doctrine Command

TRANSCOM Transportation Command

TTPI Trust Territory of the Pacific Islands

typ typical

UCMJ Uniform Code of Military Justice

UHF ultra high frequency

UN United Nations

unconventional warfare A broad spectrum of military and paramilitary operations conducted in enemy–held, enemy–controlled, or politically sensitive territory. Unconventional warfare includes, but is not limited to, the interrelated fields of guerrilla warfare, evasion and escape, subversion, sabotage, and other operations of a low visibility, covert, or clandestine nature. These interrelated aspects of unconventional warfare may be prosecuted singly or collectively by predominantly indigenous personnel, usually supported and directed in varying degrees by an external source or sources during all conditions of war and peace. Also called UW.

United States Agency for International Development The agency within the Department of State that is primarily concerned with developmental assistance and HCA. It supervises and gives general direction on all nonmilitary assistance programs under the Foreign Assistance Act of 1961, PL 480 (Food for Peace) and similar legislation. It administers HCA programs in conjunction with the US Department of Agriculture. Also called USAID.

United States Army Personnel Exchange Program A program developed to foster professional military relationships between the US Army and armies of other nations. Personnel exchange program officers are fully integrated into the host country's army, usually performing the same duties that a member of the host army would perform. These officers fill authorized positions either in the host country or the US. Also called PEP.

UNAAF Unified Action Armed Forces

US United States

USACE United States Army Corps of Engineers

USAEHSC United States Army Engineering and Housing Support Center

USAID United States Agency for International Development

USAR United States Army Reserve

USCG United States Coast Guard

USCINCPAC Commander in Chief, US Pacific Command

USIA United States Information Agency

USSOUTHCOM United States Southern Command

UW unconventional warfare

VHF very high frequency

VIP very important person

war Sustained use of armed force between nations or organized groups within a nation involving regular and irregular forces in a series of connected battles and campaigns to achieve vital national objectives. War may be limited, with some self-imposed restraints on resources or objectives. War may be general, with the total resources of a nation or nations employed, and the national survival of a belligerent at stake.

WES Waterways Experiment Station

number

@ at

References

SOURCES USED

These are the sources quoted or paraphrased in this publication.

Joint and Multiservice Publications

AR 12–15. *Joint Security Assistance Training (JSAT) Regulation (AFR 50-29; SECNAVINST 4950-4)*. 28 Feb 1990.

FM 100–20. *Military Operations in Low Intensity Conflict (AFP 3-20).* 5 Dec 1990.

JCS Publication 1–02. *Department of Defense Dictionary of Military and Associated Terms.* 1 Dec 1989.

Army Publications

AR 12–7. *Technical Assistance Field Teams (TAFT) and Technical Assistance Teams (TAT).* 15 Feb 1979.

AR 37-47. *Contingency Funds of the Secretary of the Army.* 15 Jan 1990.

AR 200–1. *Environmental Protection and Enhancement. 23* Apr 1990.

AR 350–9. *Oversea Deployment Training (ODT).* 20 Oct 1988.

AR 380–10. *(C) Policy for Disclosure of Information, Visits, and Accreditation of Foreign Nationals.* 29 Jul 1988.

AR 500–51. *Emergency Employment of Army and Other Resources—Support to Civilian Law Enforcement.* 1 Jul 1983.

AR 500–60. *Disaster Relief.* 1 Aug 1981.

AR 525–12. *Noncombatant Evacuation.* 28 May 1973.

AR 525–13. *The Army Terrorism Counteraction Program.* 4 Jan 1988.

AR 614–10. *US Army Personnel Exchange Program with Armies of Other Nations.* 19 May 77.

AR 700–128. *Prime Power Program.* 5 Jul 1985.

ER 500–1–1. *Natural Disaster Procedures.* 21 Dec 1983.

FM 31-82. *Base Development.* 3 Jun 1971.

FM 90-8. *Counterguerrilla Operations. 29* Aug 1986.

FORSCOM Regulation 55-1. *Unit Movement Planning.* 1 May 1989

TOE 05500LA00. *Battalion Headquarters Team.* 1 Oct 1989.

TOE 05510LA00. *Fire Fighting Headquarters Team.* 1 Oct 1989.

TOE 05510LB00. *Fire Truck Team.* 1 Oct 1989.

TOE 05510LC00. *Water Truck Team.* 1 Oct 1989.

TOE 05510LD00. *Brush Fire Team.* 1 Oct 1989.

TOE 05510LE00. *Aviation Fire/Crash Rescue Team.* 1 0ct 1989.

TOE 05520LA00. *Forestry Team.* 1 Oct 1989.

TOE 05520LC00. *Quarry Team, 75 tph.* 1 Oct 1989.

TOE 05520LE00. *Well Drilling Team, 1,500'.* 1 0ct 1989.

TOE 05530LA00. *Diving Control and Support Detachment.* 1 0ct 1989.

TOE 05530LC00. *Lightweight Diving Team.* 1 Oct 1989.

TOE 05530LF00. *Real Estate Team.* 1 Oct 1989.

TOE 05530LH00. *Utility (4000) Team.* 1 Oct 1989.

TOE 05530LJ00. *Power Plant Operations and Maintenance Team.* 1 Oct 1989.
TOE 05530LK00. *Power Line Team.* 1 Oct 1989.
TOE 05540LA00. *Topographic Planning and Control Team.* 1 Oct 1989.
TOE 05540LB00. *Survey Team.* 1 Oct 1989.
TOE 05540LF00. *Terrain Team (Division).* 1 Oct 1989.
TOE 05540LG00. *Map Reproduction Squad.* 1 Oct 1989.
TOE 05540LH00. *Technical Control Section.* 1 Oct 1989.
TOE 05540LI00. *Terrain Team (Control).* 1 Oct 1989.
TOE 05540LJ00. *Platoon Headquarters.* 1 Oct 1989.
TOE 05540LM00. *Topographic Maintenance Team C.* 1 Oct 1989.
TOE 05550LB00. *Dredge Team, Seagoing Hopper.* 1 Oct 1989.
TOE 05580LA00. *Light Tactical Raft Team.* 1 Oct 1989.
TOE 05580LB00. *Foot Bridge Team.* 1 Oct 1989.
TOE 05580LC00. *Assault Boat Team.* 1 Oct 1989.
Security Engineering Manual. Omaha District, US Army Corps of Engineers. Jan 1990.

Nonmilitary Publications

Public Law 79–264. *United Nations Participation Act of 1945.*
Public Law 84-99. *Emergency Flood Control Work.*
Public Law 93–288. *Disaster Relief Act of 1974.*

DOCUMENTS NEEDED

These documents must be available to the intended users of this publication.

Joint and Multiservice Publications

FM 5–166. *Well Drilling Operations (AFR 85-23).* 30 Jun 1975.

Army Publications

FM 5–100. *Engineer Combat Operations.* 22 Nov 1988.
FM 5–101. *Mobility.* 23 Jan 1985.
FM 5–102. *Countermobility.* 14 Mar 1985.
FM 5–103. *Survivability.* 10 Jun 1985.
FM 5–104. *General Engineering.* 12 Nov 1986.
FM 5–105. *Topographic Operations.* 9 Sep 1987.
FM 5–116. *Engineer Operations: Echelons Above Corps.* 7 Mar 1989.
FM 8–42. *Medical Operations in Low Intensity Conflict.* 4 Dec 1990.
FM 100–5. *Operations.* 5 May 1986.
FM 100–15. *Corps Operations.* 13 Sep 1989.
FM 100–37. *Terrorism Counteraction.* 24 Jul 1987.
FM 101–5. *Staff Organization and Operations.* 25 May 1984.
FM 101–5–1. *Operational Terms and Symbols.* 21 Oct 1985.
TC 19–16. *Countering Terrorism on US Army Installations.* 25 Apr 1983.
TM 5–301–1. *Army Facilities Components System—Planning (Temperate).* 27 Jun 1986.
TM 5-301-2. *Army Facilities Components System—Planning (Tropical).* 27 Jun 1986.
TM 5-301-3. *Army Facilities Components System—Planning (Frigid).* 27 Jun 1986.
TM 5–301-4. *Army Facilities Components System—Planning (Desert).* 27 Jun 1986.

TM 5-302-1. Army *Facilities Components System: Design.* 28 Sep 73. (With change 1, Sep 1977; change 2, May 1979; change 3, Mar 1982; and change 4, Dec 1985.)
TM 5-302-2. *Army Facilities Components System: Design.* 28 Sep 73. (With change 1, Sep 1977; change 2, May 1979; change 3, Mar 1982; and change 4, Dec 1985.)
TM 5-302-3. *Army Facilities Components System: Design.* 28 Sep 73. (With change 1, Sep 1977; change 2, May 1979; change 3, Mar 1982; and change 4, Dec 1985.)
TM 5-302-4. *Army Facilities Components System: Design.* 28 Sep 73. (With change 1, Sep *1977;* change 2, May 1979; change 3, Mar 1982; and change 4, Dec 1985.)
TM 5-302-5. *Army Facilities Components System: Design.* 28 Sep 73. (With change 1, Sep 1977; change 2, May 1979; change 3, Mar 1982; and change 4, Dec 1985.)
TM 5–303. *Army Facilities Components System—Logistic Data and Bills of Material.* 1 Jun 1986.
TM 5–304. *Army Facilities Components System User Guide.* 1 Oct 1979.

READINGS RECOMMENDED

These readings contain relevant supplemental information.

Joint and Multiservice Publications

DOD 2000. 12–H. *Protection of DOD Personnel Against Terrorist Acts.* 20 Apr 1983.
JCS Publication 0–2. *Unified Action Armed Forces (UNAAF).* 1 Dec 1986. (With change 1, Apr 1989.)
Jnt Publication 5–02.4. *Joint Chief of Staff, Joint Operations Planning System, Volume IV (Crisis Action Procedures).* 8 Jul 88. (With change 1, Jan 1989; change 2, May 1989.)
FM 20–150. *National Search and Rescue Manual (AFM 64-2; COMDTINST M–16130.2).* 1 Nov 1986.

Army Publications

AR 10–5. *Department of the Army.* 1 Dec 1980.
AR 12–1. *Security Assistance—Policy, Objectives, and Responsibilities.* 7 Oct 1988.
AR 70–1. *Systems Acquisition Policy and Procedures.* 10 Oct 1988.
AR 70–9. *Army Research Information Systems and Report.* 1 May 1981.
AR 70-38. *Research, Development, Test and Evaluation of Materiel for Extreme Climatic Conditions.* 1 Aug 1979. (With change 1, Sep 1979.)
AR 71–9. *Materiel Objectives and Requirements.* 20 Feb 1987.
AR 115–11. *Army Topography.* 1 Mar 1980.
AR 210–21. *Ranges and Training Areas.* 1 Apr 1982.
AR 405–1. *Reimbursement for General Services Administration Space, Services and Facilities Outside the National Capital Area.* 1 Sep 1983.
AR 405–10. *Acquisition of Real Property and Interests Therein.* 14 May 1970. (With change 1, Apr 1971; and change 2, Jul 1974.)
AR 420–10. *Management of Installation Directorates of Engineering and Housing.* 2 Jul 1987.
AR 500–1. *Aircraft Piracy Emergencies. 6* Oct 1972.
AR 500–50. *Civil Disturbances.* 21 Apr 1972.
AR 500–70. *Military Support of Civil Defense.* 1 Oct 1982.
AR 700–127. *Integrated Logistic Support.* 16 Dec 1986. (With change 1, Mar 1988; and change *2,* Jul 1990.)
DA Pamphlet 420-2. *Management of Fire Prevention and Protection Programs, 28* Jan 1960.

EM 385–1–1. *Safety and Health Requirements Manual.* Apr 1981. (Revised Oct 1987, with change 1, Apr 1989.)
FM 5–33. *Terrain Analysis.* 11 Jul 1990.
FM 5–480. *Port Construction and Repair.* 12 Dec 1990.
FM 20–11–1. *Military Diving (Volume 1).* 20 Sep 1990.
FM 31–20. *Doctrine for Special Forces Operations.* 20 Apr 1990.
FM 33–1. *Psychological Operations.* 31 Jul 1987.
FM 41–10. *Civil Affairs Operations.* 17 Dec 1985.
FM 46–1. *Public Affairs.* 11 Apr 1986.
FM 55–50. *Army Water Transport Operations.* 7 Jun 1985. (With change 1, Sep 1987.)
FM 55–60. *Army Terminal Operations.* 18 May 1987.
FM 90–13. *River Crossing Operations.* 1 Nov 1978.
TM 5–300. *Real Estate Operations in Oversea Commands.* 10 Dec 1958.
TM 5–315. *Firefighting and Rescue Procedures in Theaters of Operations.* 20 Apr 1971.
TM 5–330. *Planning and Design of Roads, Airbases, and Heliports in the Theater of Operations.* 6 Sep 1968. (With change 1, Sep 1970.)
TM 5–332. *Pits and Quarries.* 15 Dec 1967.
TM 5–342. *Logging and Sawmill Operation.* 23 Aug 1972.
TM 5–343. *Military Petroleum Pipeline Systems.* 20 Feb 1969. (With change 1, May 1973.)
TM 5–610. *Preventive Maintenance: Facilities Engineering, Building, and Structures.* 1 Nov 1979.
TM 5–683. *Facilities Engineering: Electrical Interior Facilities.* 2 Mar 1972.
TM 5–684. *Facilities Engineering: Electrical Exterior Facilities.* 20 Apr 1979.

Defense Technical Information Center (DTIC) Documents

DTIC Document # 385 315L. Case *Studies in the Termination of Internal Revolutionary Conflict.*
DTIC Document # A106 587. *US Strategy Needed for Water Supply. Assistance to Developing Countries (Report to Congress).*
DTIC Document # A185 972. *Operational Considerations for Military Involvement in LIC.*
DTIC Document # A185 973. *Logistical Considerations on LIC.*
DTIC Document # A209 047. *Peacekeeping Tactics, Techniques, and Procedures.*
DTIC Document # B110 687. *Interdiction of Illegal Drug Traffic-US Army Support to Civil Authority.*
DTIC Document # B123 622. *The Use of Conventional Military Forces for Drug Interdiction.*

Index

Made in the USA
Monee, IL
04 November 2026

67377586R00108